Zeenat

Manji

OIL AND ISLAM
The Ticking Bomb

OIL AND ISLAM
The Ticking Bomb

Farid A. Khavari, Ph.D.

ROUNDTABLE
—— Publishing, Inc. ——
Malibu, CA

Library of Congress Cataloging-in- Publication Data

Khavari, Farid A

    Oil and Islam: the ticking bomb / Farid A. Khavari

      p. cm.

    Includes bibliographical references.

    ISBN 0-915677-55-5 : 19.95

    1.Middle East-Politics and government-1990—   1.Title.

DS63. 1.K49   1991

956.05-dc20   90-28134

                                   CIP

FIRST EDITION

10 9 8 7 6 5 4 3 2 1

Table of Contents

This book is dedicated to the memories of my father, Noorullah, and the countless men and women who were killed by the aggression and intolerance of the Khomeini and Saddam Hussein regimes because of their steadfast beliefs.

I also dedicate this book to my mother, Banoo, who, like tens of thousands of her countrymen and women, resisted the barbaric actions of these regimes.

PREFACE

These words were written prior to Iraq's invasion of Kuwait on August 2, 1990–not as a result of it. Saddam Hussein's invasion of Kuwait simply underscored the thesis of this book. The book was updated slightly to reflect those events in the Middle East, which are a harbinger of things to come in this troubled, and troublesome, region.

Crude oil is the life of the industrialized nations. With the increasing dependency on this vital natural resource, the prosperity and very way of life of these countries are completely reliant on its availability. Equally important from the standpoint of the world economy, crude oil must be available at affordable prices.

The invasion of Kuwait by Iraq not only underscores the vulnerability of the West, but also dispels any illusion that there will be any more cheap oil.

Over sixty-six percent of the world's crude oil reserves are in the Middle East. Over eighty-four percent of all oil which is available for export stems from this critical region. And while fundamentalist Islam is gaining power, this region is becoming increasingly hostile to the United States and the West.

U.S. policy in the Middle East has been directed in the past toward establishing and maintaining governments best serving Western interests, with little or no regard to how these governments might serve the needs of their own people. In so doing, America has created monster after monster, with the late Shah of Iran, the late Ayatollah Khomeini, and the current nemesis Saddam Hussein the most prominent of them.

In fact, one need not fear the monsters so much as the short-sightedness and ineptitude of the U.S. Middle East policy. The present dangerous situation involving Iraq is, at worst, short term in nature, however drawn out it must seem from the front. The real danger lies ahead, unless U.S. policy can somehow resolve the problems connected with the Middle East and the supply of crude oil for the long term.

There is no doubt that crude oil, and Middle Eastern crude oil, will be essential to the industrialized world for decades to come, regardless of how effective U.S. Energy policy (which is virtually nonexistent) may be. The great task will still be to keep crude oil available at affordable prices. However, there is no guarantee that this can be achieved, unless serious, thoughtful long-term policies are contemplated...and enacted.

Therefore, the primary goal in this book will be to demonstrate the complexity of the background of the problems in the Middle East, and to call attention to potential threats to the prosperity and very survival of industrial civilization–threats which the industrialized nations of the world, particularly the United States of America, cannot afford to ignore. Strategies and policies will be proposed which would not only secure the interests of the industrialized nations, but which would also best serve the interests of the peoples of the Middle East.

OIL AND ISLAM
The Ticking Bomb

1

THE TICKING BOMB

*T*he West, and particularly the United States, owes a debt of gratitude to both the Ayatollah Khomeini and Saddam Hussein. After all, they forestalled an oil crisis for nearly a decade. The industrialized world owes a further gratitude to Saddam Hussein: he could have taken over Saudi Arabia, too. Nevertheless, he overran the tiny country of Kuwait on August 2, 1990, and confronted the West with undeniable proof of the fragility of its economic well-being.

Had Iran not bungled the war with Iraq, or if Khomeini had had the wisdom to accept the peace desperately offered by the Iraqis, the Saudis and the other Arab nations in 1982, the Middle East would have been a very different place today. Iran would have controlled the entire region–and of the Organization of Petroleum Exporting Countries (OPEC) as well–truly a global superpower. Instead, Khomeini's obsession with destroying Saddam Hussein practically destroyed Iran. Khomeini tasted defeat before he died, and the stage was set for Saddam Hussein to take advantage of the opportunities in oil and real estate in the Middle East.

Had Hussein invaded Saudi Arabia on August 3, before

the West could mobilize its military might, he would control 44% of the exportable oil in the world, rather than the mere 22% he has with Iraq and Kuwait. In that scenario, the West would not have had a viable military option, and would have been at the mercy of one man.

Despite the West's feeble efforts to conserve energy and find alternatives to oil since the early 1970s, crude oil's share of total energy consumption remains constant, while total demand for energy continues to climb. The United States' dependency on Middle Eastern oil has not abated; it has increased at an alarming rate.

There is no magic solution to the West's dependency on foreign oil. Some researchers have recommended letting free market forces rectify the situation[1]. They argue that without government interference, the laws of supply and demand would somehow deliver substitutes for crude oil. The extent of the naivete expressed in this view will become clearer later on as this question is explored. If the technology existed to replace Middle Eastern oil with alternative energy sources, it would take years to implement it. Even if the technology and the determination to implement it existed, it could not resolve a crisis before the economies of the industrialized world collapsed.

Meanwhile, the United States has no coherent energy policy. The technology to replace crude oil in sufficient quantity to eliminate the dependency on foreign oil is undeveloped. And where will the hundreds of billions of dollars come from to finance such a project?

Clearly, the supply of oil from the Middle East must remain available and affordable for the foreseeable future, if industrialized society is to survive and have time to wean itself of imported oil. The only security the West can have is stability in the Middle East.

Iraqi military forces occupy Kuwait, a multinational force–largely American–is rapidly building up to confront them, and Iraq and Kuwait are under trade blockade. Without underestimating the significance of these events and whatever the outcome of Saddam Hussein's adventure, this spectacle must not divert our attention from the long-term threat to the interests of the industrialized world. This threat began long before Saddam Hussein came into power and will remain when he is no longer on the scene. This threat is exacerbated by the presence of Western military power in the Middle East, yet will not respond to military force.

The most serious threat to stability in the Middle East and to the economic future of the industrialized world is the growing influence of fundamentalist Islam–radical, revolutionary fundamentalist Islam–in Middle Eastern politics.

Foreign powers, especially Great Britain and, the United States in recent years, have traditionally meddled in the politics and economics of the Middle East, attempting to shape events in the region to suit their own purposes. The result of this interference is, essentially, the history of the region in this century.

Nevertheless, such actions are facts of life. Ironically, the fact that fundamental Islam is an inexorably growing threat to the world economy–and to world peace–is largely due to the efforts of the West, with special emphasis on the United States of America.

The ticking bomb must be defused, and this cannot happen without external influences. The threat must be understood, the implications grasped, and appropriate action taken. Fortunately, there is no need to agonize over ethics in this situation. In this case, the interests of both the industrialized world and the people of the Middle East themselves coincide. For the industrialized nations, the stakes are their economies and their way of life. For millions of good, decent people in the Middle East, the stakes are life and death.

The recent history of U.S. involvement in the Middle East would lead one to believe that the storehouse of American solutions is Pandora's Box.

America's first approach to controlling the Middle East was to simply establish U.S. puppets in the countries which were important to U.S. interests. Iran's geostrategic position made it an especially compelling stage, with the late Shah Mohammed Reza Pahlavi as the star puppet, surrounded by the kings and sheikhs ruling the other countries in the area. The Shah served U.S. interests loyally and vigorously. He spent billions on American weaponry and built the sixth most powerful military force on earth at his country's expense to be the enforcer of U.S. policy in the region. The Shah was a bulwark of stability in the Middle East as far as Western interests were concerned. There were, however, those who thought he was too hard on the folks at home. Then he began to get a little bit headstrong economically, and it was time for the U.S. to take him off the stage.

American policy took a sudden turn to become one of shortsighted ignorance: it brought Ayatollah Khomeini's Islamic Revolution to power in Iran, and thus unleashed the gravest threat to the security of the industrialized world since World War II.

It didn't take long for the West to realize that Khomeini was dangerous. He quickly began to spread his revolutionary fervor throughout the region. Fortunately for the West, there was Saddam Hussein, who didn't require much encouragement to attack Iran. Hussein polarized the Arabs against Khomeini, who threatened the very lives of the ruling class throughout the Middle East. American policy shifted from one of ignorance to one of choosing the lesser of two evils. Saddam Hussein became the next champion of Western interests as he waged an eight-year war against the Islamic Republic.

But when the war was over, here was an arrogant victor, in charge of a ruined economy and a million men at arms with a lot of time and bullets on their hands. Even after weeks of an Iraqi military build-up on the Kuwaiti border, no one believed Saddam Hussein's threats to invade Kuwait until it actually happened. By invading, occupying and later annexing Kuwait to Iraq, Hussein demonstrated that he may have further territorial ambitions. But the most menacing aspect of his actions is that the economy of the industrialized world was at stake.

The U.S. then imposed an economic blockade and began a crash build-up of military forces across the border in Saudi Arabia in August of 1990. There is no doubt that Saddam Hussein's days are numbered. But what comes next?

Every day of foreign military presence in the region fans the fires of the Islamic fundamentalists by its very existence. Whether the West defeats Hussein, or the other way around, one of fundamentalist Islam's enemies is defeated, and the other is weakened. Whatever the outcome of the situation, it paves the way for Islamic Revolution to sweep across the region, with the Islamic Republic of Iran at the center of the movement.

America can ill-afford to repeat its past policies in the Middle East, or to ignore the trends in such a vital area. What is needed is a long-term strategy to bring peace and stability to the region, one which also addresses the interests of the region's population rather than simply those of the West. If the United States fails to act with the western industrialized nations to support, and establish, governments in the region which serve the interests of the world and of their own

citizens, the next monster in the Middle East will be one
which is immune to military response: fundamentalist Islam.

2

SADDAM HUSSEIN:

The Lesser Evil?

Not so long ago, Iraq's president Saddam Hussein was the hero of the western nations, fighting Ayatollah Khomeini's Islamic Republic and preventing the spread of radical Islamic influence throughout the Persian Gulf region. With his ruthless ambition and appalling barbarism, Hussein was not the sort of hero the West was fond of; it was just that, when Khomeini became uncontrollable, Saddam Hussein, with his desire to eliminate Khomeini and the Islamic Republic, seemed the lesser of two evils. Saddam Hussein abdicated his very short pedestal when he invaded Kuwait.

The West has long been wary of Saddam Hussein, and rightfully so. Not content with the cornucopia of high-tech weapons available to him from eager arms dealers of the East and West, Hussein aspired to the power afforded by exotic instruments of mass destruction. His program to develop thermonuclear devices was postponed, if not thwarted completely, by an Israeli air strike on his nuclear plant in 1981.

The war with Iran was the proving ground for Hussein's chemical weapons: it proved that he had chemical weapons and the means to deliver them, that they worked, and that Hussein was not afraid to use them, against Iran, or even Kurds in Iraq.

Still, the fear of Khomeini's radical actions led the West to ignore the potential threat of Hussein, as long as his actions served to suppress the spread of Khomeini's Islamic Revolution. Financial and military aid, both directly and indirectly, was heaped upon Iraq to bolster Saddam Hussein's efforts to wipe out the very real threat of Khomeini. Eight years later, when Hussein was victorious, he was left with a devastated economy, tremendous debt, an army of a million men (out of a total population of only 17 million), and frustrated ambitions. It should have appeared obvious to the West that Iraq was a war looking to happen.

Although out of the limelight, Hussein was busy rebuilding his arsenal. He regained world attention by hanging Iranian-born British journalist Farzad Bozoft, in March 1990. Then came the interception of a shipment of capacitors designed as components of triggering devices for nuclear weapons, enroute to Iraq, and of a shipment of massive steel tubes apparently intended for the construction of a "supergun." Suddenly, Saddam Hussein became an evil force with which to reckon.

With the invasion of Kuwait, Saddam Hussein reached the nadir of his relationship with the West. President George Bush of the United States declared that Hussein must be removed from power or even eliminated. Bush initiated a total trade blockade against Iraq and commenced the largest military build-up since World War II to confront him. The minimum goal of the West was to force Iraq to withdraw from Kuwait and restore *status quo ante*.

There is no doubt that Saddam Hussein has ambitions which extend far beyond Kuwait or the leadership of the Arab world. The oil-dependent western industrialized nations reacted vigorously to the invasion of Kuwait, realizing that they could not allow the fate of the world economy to rest in the hands of the brutal megalomaniacal Hussein.

However, Saddam Hussein is not the only villain in the real story of the making of this monster. He could not have achieved his power without the help of many nations, from the West, the East and even the Third World. Among the nations who stocked Hussein's armories were Russia, France, Brazil, China, Czechoslovakia, Egypt, South Africa, Great Britain, Belgium, Italy, Sweden, West Germany...and the United States. One must not discount Hussein's own efforts to enhance his arsenal by employing various foreign experts in his weapons development programs. Most notable among these were Dr. Gerald Bull, a Canadian scientist, and retired Brazilian Air Force Brigadier General Hugo Piva, the former director of Brazil's aerospace agency, *Centro Tecnico Aerospacial*. Iraq became a great recycler of abandoned weapons projects from around the world.

Iraq needed the means to deliver its warheads to the heart of Iran. The search for such technology led to a Canadian scientist who was a veteran designer of military hardware. Dr. Gerald Bull had been involved in the 1951 joint U.S.-Canadian Project HARP (High Altitude Research Project), which was intended to create a supercannon capable of launching a projectile into space. Three sixteen-inch diameter supercannons were built, but HARP was superseded by other advances in rocketry. Bull then formed Space Research Corporation in Quebec, and offered his services to the world. When the CIA learned of this, Bull was imprisoned for three months, and subsequently relocated his company in Brussels, Belgium.

During the Iran-Iraq war, Dr. Bull met Iraq's Minister of Industry, General Hussein Kamel, with whom he developed a close friendship. Kamel was an influential member of Iraq's Baath Party and was responsible for Iraq's weapons development program.

Dr. Bull's involvement in Iraq's weapons program resulted in Space Research Corporation placing orders with British firms Sheffield Forge Masters and Walter Somers Ltd. to manufacture different sizes of large metal tubes for Iraq's supergun. Sheffield was to fabricate the larger tubes, and Somers the smaller ones. Meanwhile, Bull transferred the operation of his company to his sons, Michael and Stephen, and concentrated wholly on developing the supergun.

Space Research ordered fifty-two large steel tubes from Sheffield, at a cost of $1.7 million. Forty-four of the tubes had already been delivered when the last eight were intercepted at a British port. By then Iraq already had the materials to build at least five superguns, and the loss of the eight tubes was insignificant.

A Muslim worker at Sheffield learned of the destination of the tubes, and reported this information to the Iranian Embassy in London. The embassy contacted Tehran, who notified MOSSAD, the Israeli secret service. On December 22, 1989, as Dr. Bull left his apartment in Brussels he was shot three times in the head. Then MOSSAD informed the British Secret Service, who took no action until the hanging of Farzad Bozoft.

Even on a grander scale than that of the supergun project was Iraq's aerospace program, staffed by engineers from West Germany, Egypt, Argentina and Brazil. The stars of this show were approximately 40 Brazilians, led by Hugo Piva, a noted Brazilian scientist.

In Brazil, Piva coordinated efforts to develop the Piranha air-to-air missile, a copy of the American Sidewinder, for deployment under the wings of jet fighter planes. Just as the Piranha was nearing production, Brazil scrapped the project in late 1988 due to financial problems. Piva formed a private firm and sought to profit from his previous achievements. In early 1989, Piva found a customer. He and his group were hired to produce a version of the Piranha in Baghdad. Speculations abounded that Piva was also involved in a project to modify Soviet SCUD ballistic missiles, perhaps to accommodate nuclear warheads, or to produce a new long-range ballistic missile.

Saddam Hussein did not build the world's fourth-largest military force without a lot of help, not only in the form of weapons and technology from many sources, but of financial aid as well. Ironically, in order to provide trade credit to the Iraqis, the United States government first had to circumvent its own policy. In spite of the political repression, brutality and terrorism of Saddam Hussein's regime, the U.S. in 1982 declared that Iraq was no longer a terrorist state.

After the cease-fire in the Iran-Iraq war, the Iraqi people demanded improvement of their economic lot, which created a serious dilemma for Saddam Hussein. Due to economic havoc from eight years of war, Iraq was in desperate need of foreign loans, but no country was willing to extend any more help. Even France and Italy, Iraq's two major trading partners, were disillusioned and decided not to grant further credit.

However glorious its military might, Iraq was destitute, with over $70 billion in foreign debt, of which some $40 billion was owed to the West and to the U.S.S.R., with most of the balance owed to Arab nations in the Persian Gulf region. Along with a bankrupt economy and shrinking oil income, Saddam Hussein was saddled with the huge cost of maintaining his victorious, and expensive, military machine, at a cost of over $5 billion per year. Compared with the U.S. defense budget this may not seem like a lot. But when one considers Iraq's Gross National Product of $52 billion on one hand, and the interest on its foreign debt on the other, it is indeed a heavy load for an impoverished country of only 17 million people. It was difficult for Iraq to bear the burden for long. Conquest became a more and more attractive alternative to Hussein, both as a source of revenue and to distract the attention of his people from their grim surroundings. Iraq crossed the border into Kuwait and simply took over. Saudi Arabia might well have been next had not the United States responded immediately with military support for that country.

Saddam Hussein developed his taste for politics and militarism early in life. He was born in 1937, and became orphaned before his first birthday. The uncle who raised him was an army officer who railed against Britain's control of Iraq's puppet monarch, King Faisal (incidentally, the cousin of Jordan's King Hussein). The uncle was imprisoned in 1941 for his participation in an attempted anti-British coup.

Saddam Hussein began school at age nine, and didn't do well. He was rejected by the Baghdad Military Academy because of poor grades, which was a hard blow to a lad with military ambitions. Hussein's own experience with violence as a political tool came when, as a student, he joined the radical anti-Western, pan-Arab socialist underground movement. The party assigned him to a squad whose mission was to assassinate Iraq's military ruler, Abdul Karim Kassem. The squad raked Kassem's passing car with machine-gun fire, but missed their mark. Kassem's bodyguard dispatched several of the assailants, and Hussein escaped with a bullet in his left leg. He made his way to Syria, where he was captured and imprisoned. Egyptian President Gamal Abdul Nasser, sympathetic to the Baath Party's pan-Arab designs, had Hussein transferred to Egypt.

At the age of 25, Hussein took up the study of law, but his heart wasn't in it. He returned to Baghdad in 1963 and organized a militia for the Baath Party. The Baath Party finally seized power permanently in 1968. After playing a strong role in the government for a decade, Saddam Hussein assumed power in 1979 upon the retirement of General Ahmed Hasan al-Bakr. Hussein promptly made himself a full general. One of his first acts as the new ruler was to issue an order for the execution of 21 cabinet members.

Saddam is not a very popular leader among his people, or his army. During the invasion of Kuwait, he executed 120 Iraqi officers and soldiers who refused to attack another Muslim country. Many of the invading Iraqi soldiers believed that Kuwait had been attacked and that they were mobilizing to defend their fellow Arabs.

Hussein's power is based on fear, and, ironically, he lives in constant fear of assassination and coup. He can't eat a meal without someone first sampling the food; this precaution once saved his life in 1989. He wears a bulletproof vest, never travels a pre-planned route, changes his guards daily, and never sleeps in the same place two nights in a row. His fears are well justified. In June, 1990, about thirty Iraqi military officers, including three generals (Maher Abdul-Rashid, Hamid Shaban, and Lice Mohammed) were imprisoned for plotting a coup. A few weeks prior to the invasion of Kuwait, a group of officers from the al-Shagat barracks was arrested for planning to assassinate Saddam during a military ceremony. Again on August 19, another plot was uncovered.

Saddam Hussein has kept his grip on power by systematically eradicating any encountered opposition. In 1982, when

Iraq's prospects in the war with Iran were grim, and Ayatollah Khomeini demanded Hussein's ouster as a prerequisite for peace, Iraqi Health Minister Riaz Abrahim suggested in a cabinet meeting that Saddam might resign for a short while. Saddam calmly invited Abrahim into the next room, drew his pistol and shot him on the spot. Hussein's opponents are usually dispatched more spectacularly.

In 1971, when Saddam was Vice President, General Hordan Takriti, who was Deputy Prime Minister, Deputy of Commander in Chief of the Iraqi Army and a member of the Revolutionary Command Council was removed and exiled to Kuwait. There, he was machine-gunned in his hospital bed. When Saddam became president in 1979, he ordered summary executions of army officers whose loyalty he doubted. Several generals are taken up in exploding helicopters. Even family members aren't left out. Defense Minister Adnan Khirollah, Saddam's cousin/brother-in-law (Saddam married his cousin, the daughter of the uncle who raised him, Khirollah's sister), was vaporized in a helicopter shortly after protesting Saddam's plans to take a second wife.

If Saddam Hussein were killed, chances are that the Iraqi army would not fight for him any more. When the late Shah of Iran left his country, one of the world's great armies fell into disarray and surrendered to a ragtag bunch of thugs. Like the Shah, Saddam is a one-man show. Without Saddam to drive it, it is unlikely that the Iraqi army would sacrifice itself to the combined military might of the West.

Much has been written about the enormous power of Saddam Hussein's war-tempered military machine. His army is said to number a million men; the Shah of Iran at the height of his power had a force of no more than 350,000. Iraq is reported to have 5,500 main battle tanks, 3,700 artillery rocket launchers, over 500 combat aircraft and 160 armed helicopters, as well as chemical weapons and possibly nuclear devices. Clearly such a force poses a grave threat to the entire Persian Gulf region.

Former U.S. Secretary of Defense Casper W. Weinberger stated that "Iraq's army is large but was never able to do more than hold off Iran. It was our successful protection of Kuwait's tankers that convinced the Iranians they could not win, and they gave up."[2] Still, one cannot discount the military threat of Saddam Hussein. Nor can one overlook the powerful deterrents of nearly 14,000 western hostages shielding strategic military and industrial sites, and the vulnerability of Middle Eastern oil installations.

Another indication of the limitations of Saddam's military might is the fact that he did not invade Saudi Arabia. This would have secured a dominant position in the Persian Gulf, and as a result, in the world oil market. He could have placed charges on all of the oil wells and effectively neutralized any threat of retaliation. Instead, he stopped at the border, allowing western forces to mobilize. And, while the western response was admirably quick, he could have attacked when the first forces arrived in Saudi Arabia. Every day that he delayed allowed the opposing forces–and the certainty of his defeat–to grow stronger. Why didn't Hussein take Saudi Arabia and consolidate his power when he had the chance?

Commentators and analysts say that the West caught Saddam off guard by responding with troops. They say that Saddam didn't imagine the Americans would fight him, that the thousands of hostages and lack of American resolve would protect him from armed response. Of course, the Americans preferred economic sanctions, diplomacy and so forth, because it takes time to mobilize. But in fact, it was the West who was caught off guard.

The real reason that Saddam did not invade Saudi Arabia, and perhaps the reason that he did invade Kuwait, is that

Saddam is afraid of his own army. While Saddam rules by fear, he is also ruled by fear. To occupy territory larger than a helpless Kuwait would require deployment of far too many tanks and troops–and loyal ones at that–for comfort. Armaments might be needed in Baghdad at any time to put down a coup.

As a ruler, Saddam Hussein combines the ambition of the late Shah of Iran with the brutality of Ayatollah Khomeini. Just as the Shah wanted to revive the glorious past in celebrating 2,500 years of Iranian monarchy at Persepolis in 1971, Hussein dreams of rebuilding the ruined splendor of the Babylonians. Like the Shah's SAVAK, Hussein's highly efficient secret police are pervasive and effective at keeping the population under control. However, when it comes to methodology of repression, Hussein surpasses Khomeini: opponents are simply executed or tortured and killed. Hussein occasionally dispenses justice by his own hand.

Like the Shah's, Saddam Hussein's regime is based on one man. Remove Hussein, and major changes could occur in Iraq, as they did in Iran when the Shah was forced from power. It is important to note the difference in the Islamic Republic of Iran, which is based on ideology and where little changed after Khomeini died.

Saddam Hussein was not only a bitter enemy of Khomeini, but of the Shah as well. Unlike Khomeini, the Shah knew how to deal with Saddam Hussein and to keep him in his place. With little effort, the Shah forced the Agreement of Algiers upon Hussein in 1975. This agreement gave Iran sovereignty over the waterways of the Persian Gulf. After the Iran-Iraq war, Hussein tried to reverse the agreement, but faced with the harsh reaction of the West to his invasion of Kuwait, Hussein was forced to cede disputed territory to Iran, and retract his demands for a reversal of the Agreement of Algiers.

Although Saddam Hussein is generally acknowledged as the victor in the eight-year war with Iran, it may be too soon to declare the ultimate winner. Iran suffered hundreds of thousands of casualties, and as many as 150,000 Iraqis were killed. Iraq was buried in debt with its economy destroyed; while Iran had no friends to lend it money, and thus finished the war with no new debt. Iraq's victory may have been illusory.

In time, many words will be written about the rise and fall of Saddam Hussein. For now, his fate appears to be in the governments of the United States and the West. So far the only person who ever made Hussein back down was the Shah of Iran. At the Conference of Algiers in 1975, Saddam Hussein told the Shah, "Whatever His Majesty says, I will obey." That humble statement cost Iran dearly seven years later.

Although one must not underestimate the threat of a massively- armed madman who has demonstrated his willingness to use chemical weapons, Saddam Hussein ironically does not represent a long-term threat to world peace, if his actions can be thwarted and his ambitions tempered.

Regardless of how the United States and the West decide to deal with the crisis instigated by Hussein, four objectives must be at the center of any solution: 1)Iraq's million-man army must be disarmed, or at least be reduced to a non-threatening force; 2) Iraq's chemical weapons and the facilities to produce them must be eliminated; 3) any facilities with the potential for producing atomic weapons in Iraq must be destroyed; and 4) weapons delivery systems, including Iraq's missile technology, must be neutralized.

In realizing these objectives, the West must avoid jeopardizing the flow of oil from the Persian Gulf region. The oil fields and production facilities are extremely vulnerable.

Should the West subdue Saddam Hussein only to have these precious assets destroyed by commando raids, the cost could be the economies of the industrialized nations. After all, the whole idea behind the West's massive efforts is to protect the flow of oil *at an affordable price.* Hussein is in an excellent position to slash the jugular vein of world economy.

Saddam Hussein has his own areas of vulnerability. His cruel and repressive regime is hated by the Iraqis as much as Iran's radical theocracy is hated by the Iranians. Hussein might not have remained in power for so long had it not been for the war which he started with Iran. In fact, he owes much of his longevity to his arch-enemy Ayatollah Khomeini, who bungled Iran's chances for an early victory and then relentlessly stretched out the war at incredible cost in lives and material. The war was a two-way street: focusing Iranians' attention on an external enemy enabled Khomeini to entrench his radical regime; "winning" the war strengthened Hussein's position in Iraq.

Hussein's internal enemies include the Shi'ite Muslim majority and the Kurds. Sunni Muslims, Saddam's power base, make up only a quarter of Iraq's population of 17 million. Over half of all Iraqis are Shi'ites, and the remaining quarter are Kurds. Hussein killed a lot of Ayatollahs, and he used chemical weapons on the Kurds. The Shi'ites could be a great force in destabilizing Hussein's government from within. Should the Shi'ite majority gain power in Iraq, it would represent a tremendous boost in momentum for the Islamic fundamentalist movement in the Middle East. The largest opposition factor in Iraq today is "al-Dawa" (the Call), a radical fundamentalist Shi'ite group. However, as will be explored, the greatest mistake of all would be to replace Saddam Hussein with an Islamic theocracy.

3

SHAH MOHAMMED REZA PAHLAVI:

A Lesson From the Past

Not only was the downfall of Mohammed Reza Shah, the late Shah of Iran, one of the great tragedies of recent history, it proved to be a very big mistake for practically everyone who contributed to it. The Shah's regime was far from perfect, but the conspiracy to bring him down was disastrous. Neither the Iranian people nor the conspirators derived any benefit from the fall of the Shah, and in its wake came catastrophic consequences for practically everyone: executions, war, death, economic and political chaos, hostage-taking, hijacking and other terrorist activities. The sole beneficiaries of the fall of the Shah were a group of Iranian clergymen, the moguls of Qom.

To understand the Iranian revolution and the fall of the Shah, it is necessary to examine the realities which influenced the thoughts and actions of the Shah from the beginning of his reign in World War II, when Iran was a country in disarray, until 1978, when Mohammed Reza Shah Pahlavi was generally regarded as one of the world's most powerful and richest rulers, and finally to early 1979, which saw the beginning of his fall.

It is also necessary to understand the geography and history of Iran.

Iran is a vast country, 628,000 square miles in area, larger than the combined areas of France, Spain and Italy. Its population of approximately 51 million is made up predominantly of descendants of the Aryans. The word Iran means "Land of the Aryans".

Iran shares a long border with the Soviet Union to the north. The Persian Gulf and the Gulf of Oman form Iran's southern border. Iran is flanked by Iraq to the southwest, Afghanistan to the northeast, and Pakistan to the southeast. Its geographic position, between the Soviet Union and Africa, in the heart of the Middle East, is one of unique strategic importance in the world.

Iran is part of the Great Iranian Plateau with an average elevation of about three thousand feet above sea level. This dictates the climate, with rainfall scarce except along the Caspian Sea. Man has inhabited Iran for at least 8,000 years. Recent excavations of early civilizations reveal that the ancient Iranians had high aesthetic values. Iran's recorded history begins with the Medes of the seventh century B.C.

About 550 B.C., Cyrus the Great established the political supremacy of the Achaemenians, uniting the kingdoms of the Medes and the Persians. The next twenty years were spent building an empire which, at its peak, stretched from Central Asia to the Mediterranean Sea. His tolerance and humanitarianism made Cyrus the Great one of the most sympathetic figures of the ancient world. His humane treatment of the Jews after the conquest of Babylon was praised by writers of the Old Testament.

The next great dynasty was that of the Parthians, who battled with the Roman Empire for almost 500 years, mainly for control of the territory lying between Iran and the Mediterranean. The Parthians were followed by the Sassanians (224 - 651 A.D.) under whom Iran enjoyed to some extent the fabled glory of the past. However, weakened by the Romans, their battered armies fell to the Muslim Arabs. As a result, Islam was imposed upon Iran, and the country fell under the rule of the Caliphates. Soon a series of dynasties reasserted the independence of Iran. The most important of these were the Saffavids, the Samanids, the Buwayids, the Ghaznavids and the Saljuks.

Early in the thirteenth century, Iran was attacked by the Mongols. This destroyed the country and paralyzed the progress of its civilization for the next 150 years. When this period ended in anarchy, Iran experienced a renaissance under the Saffavid Shah Abbas the Great (1587-1628). Successive Saffavid rulers brought peace and a measure of prosperity to Iran, although leaving much to be desired.

By the end of the first decade of the twentieth century, Iran was once again torn apart by the backlash of foreign wars. Internal chaos followed, and the country's survival as an independent nation was on the verge of collapse. Iran's king, Ahmad Shah, the last of the Qajar dynasty, was an incompetent, corrupt weakling. Ahmad Shah was eventually ousted by one of his soldiers, Reza Khan. After reuniting the country and reestablishing its progress, Reza Khan proclaimed himself emperor and took the title Reza Shah Pahlavi on April 25, 1926.

Reza Shah, founder of the Pahlavi dynasty and the father of Mohammed Reza Shah, was, in fact, one of the greatest kings of Iran. He set the stage for change in Iran, turning an economically backward country into a relatively modern and progressive nation. The Iranians still believe that had Reza Shah remained in power for only another decade, he would have fulfilled his plans and transformed Iran into a modern nation, comparable to any in Western Europe. Even today, Iranians feel that if Reza Shah had been in power instead of his son, Iran would not have fallen into the hands of Khomeini, and that Iran can be rescued if only it produces another king like Reza Shah the Great.

Reza Shah's popularity was based on his strong determination to transform Iran into a modern nation, and his boldness in the execution of his plans. This latter quality was sadly lacking in his son, Mohammed Reza Shah.

British and Russian forces invaded Iran on August 23, 1941. Their primary objective was control of the trans-Iranian railroad. Secondly, they wanted to remove Reza Shah, whom they accused of sympathizing with Hitler and the Nazi party. Iran had become an area of prime strategic importance during Germany's onslaughts against Russia during World War II. The only safe and relatively rapid passage to supply Russia with war material was through the Persian Gulf.

The Allies' hostility toward Reza Shah was inspired partly by his personality and partly by their perception of his rapid modernization as a threat to the balance of power in yet another region of the world. They feared he could become another Hitler. The invasion of Iran allowed the British to get rid of him. Reza Shah's abdication followed on September 16, 1941, and he was exiled to South Africa, where he died in 1944.

The circumstances surrounding the coronation of Mohammed Reza Shah were less than enviable. He was twenty-two years old. British and Russian forces occupied Iran. Not only were the British and Russian ambassadors conspicuously absent from the coronation ceremonies, but the British inclination was to support a Qajar prince who was an officer in the Royal Navy, rather than Mohammed Reza.

Similarly, the circumstances during Mohammed Reza Shah's rule were seldom ideal. Many of the volatile situations the young Shah faced threatened his position of power, or at least brought relentless accusations of being a puppet of some superpower or another. The extent of wretchedness in Iran was due not only to its economic backwardness, but further to the interplay of several factors with which the Shah had to contend. These factors prevented Iran from escaping the vicious cycle in which it was trapped, and ultimately contributed to the Shah's downfall.

Mohammed Reza Shah's reign began shortly after the invasion by the Russians and the British. Until 1946, he concentrated on preserving Iran's economic and political integrity. The provinces of Azerbaijan, Kurdestan and Khuzestan would have gained autonomy, had the Russians not rejected the plans of British Foreign Secretary Ernest Bevin and U.S. Secretary of State James Byrnes at the Moscow Foreign Ministers' Conference of December, 1945. The Russians wanted Iran's collection of provinces intact because they expected the whole country to fall under their jurisdiction. This expectation met with disappointment, however, and finally, at President Harry S. Truman's assistance, the Russian Army withdrew from Iran in May, 1946. The stage was now set for Mohammed Reza Shah's next test, the Adventure of Azerbaijan.

This adventure was underwritten by the Russians and led by a man named Pishevari. Its primary goal was to separate Azerbaijan from Iran. A military operation led by Mohammed Reza Shah himself defeated Pishevari and returned Azerbaijan to the fold.

The Shah's next challenge was to survive the era of Dr.

Mohammed Mossadegh's premiership, when events would lead the Shah to flee for his life. Much has been written about Dr. Mossadegh. He is considered to have been a nationalist whose objective was to eliminate the pervasive economic and political influences and interferences of foreign powers, mainly those of the British, through nationalization of the British-dominated Iranian oil industry.

However, the Shah portrayed Mossadegh as pro-British. In his book, *Answer to History*, the Shah stated, "We [Reza Shah and Mohammed Reza Shah] always suspected that he [Mossadegh] was a British agent, a suspicion that his future posturing as an anti-British nationalist did not diminish."[3] Examination of Mossadegh's political decisions and his actions during his premiership, especially those concerning nationalization of the oil industry, shows that the Shah's suspicions were appropriate.

One could point to Dr. Mossadegh's deeds as evidence of his patriotism, and blame his ultimate failure on his inexperience and lack of knowledge of the oil business. Nevertheless, Mossadegh should have anticipated such a radical move as nationalizing the oil industry of a country with limited technical capabilities, and lacking access to foreign markets, could only lead to economic and political catastrophe.[4] The utter predictability of such consequences reinforces the suspicion that Mossadegh was acting deliberately on behalf of some foreign power. Certainly the Iranians did not benefit from his actions.

There is no question that the British exploited Iranian oil, and consequently the country. This began in 1901. An agreement was signed awarding the rights of exploration, production, refining, sale and transportation of all of Iran's oil to one William Knox D'Arcy, a British entrepreneur. Although revised agreements somewhat more favorable to Iran were reached in 1920 and 1933, the policy of the Anglo-Iranian Oil Company (later to be known as the British Petroleum Company) was one of continued discrimination against Iran. Britain paid considerably higher royalties to other countries in the Persian Gulf region, including Iraq and Kuwait.

The dissension between Iran and the Anglo-Iranian Oil Company reached the breaking point when American oil companies signed an equal percentage contract with Saudi Arabia. Anglo-Iranian was paying Iran less than 30%. The Iranians were further dissatisfied with the company's failure to train Iranian technicians and reduce foreign employment in Iran, the miserable wages paid to Iranian employees, and the company's investment of profits on Iranian oil in oil exploration outside of Iran.

In May, 1951, the law nationalizing the oil industry was ratified by the Iranian Parliament, and subsequently enforced by Dr. Mossadegh. The result was virtual economic suicide for Iran, whose lack of production know-how or marketing capabilities left it bereft of its primary source of income. Britain immediately protested, taking the dispute to the International Court of Justice in the Hague, withdrawing its 4,800 technicians from Abadan, and blockading Iran's ports. Iran was unable to sell any oil for three years, even at half price. Without income from oil, Iran's economy began to decay rapidly. The most Mossadegh gained by nationalizing Iran's oil industry was a few buildings, desks and typewriters. Could this have been his patriotic goal? Three years later, after Mossadegh was deposed, a face-saving agreement was reached in 1954 between a consortium of eight of the world's largest oil companies and the National Iranian Oil Company. Income from the sale of oil once again flowed into the Iranian treasury.

One cannot simply dismiss the possibility that the

nationalization of the Iranian oil industry was part of a British master plan designed to intimidate the other oil-producing countries of the Middle East and discourage any thoughts they might have with regard to nationalizing their own oil industries. Despite protestations and reprisals, no one derived greater benefit from Iran's nationalization of its oil industry than the British.

There has been considerable controversy surrounding the overthrow of Mossadegh and the Shah's return to power. The Shah believed that his popularity among the Iranian people brought this about, moreso than the support of the United States and Great Britain.[5] According to the Shah's twin sister, Ashraf Pahlavi, however, there was strong involvement right from the beginning on behalf of the Americans and the British. In her memoirs[6] she details this involvement, and attributes the Shah's return to his throne to the work of the American Central Intelligence Agency, while acknowledging the Shah's popularity.

Mossadegh brought plenty of grief to the Shah, and caused tremendous suffering for the country. Contrary to Mossadegh's avowed goal of eliminating foreign influence, his actions allowed American meddling in Iran's economic and political life in direct competition with the British. The Shah would not look back fondly on the Mossadegh era. He nearly lost his throne, and, had Mossadegh retained power, Iran might have become a communist state.

After the disastrous Mossadegh era, threats of communism, the menacing Russians, the hardships imposed on Iran by the British, and his need for U.S. support to restore his regime, by 1954 it was clear to the Shah that a close and friendly relationship with the United States was of immense importance. The Shah looked to the U.S. not only for financial assistance, but for military aid as well. This brought the Shah into disfavor with the Soviet Union. Moscow viewed American involvement in Iran as detrimental to Soviet security. Although the Soviets never took a radical step, fearing that to do so would push the Shah closer to the Americans, they did not ease their pressure on the Shah until the late 1960s when the Shah stepped up trade with the Soviet Union.

The overthrow of Mossadegh did not solve all of the Shah's problems, nor did it eliminate all of his enemies. Two attempts were made on the Shah's life, the first in February, 1949, by Fakr Arai, a member of the Tudeh Party (the Iranian Communist Party) who was also involved with an ultra-conservative fanatical religious group. The second attempt occurred on April 10, 1965, when a soldier broke into the Shah's Marble Palace and fired a machine gun toward the Shah's room. Surrounded as he was by declared enemies, it is ironic that the Shah's undoing was at the hands of his so-called friends.

Soviet criticism of Iran and the Shah coincided with disapproval of the Shah's policies by the growing radical Arab forces, notably Egyptian President Gamal Abdul Nasser (1954-70). Nasser condemned the Shah's regime as anachronistic, anti-revolutionary, anti-progressive, and being propped up by Washington to promote and nurture America's "imperialist interests" in the region.[7]

After the July, 1958, overthrow of the pro-British Hashemite monarchy of Iraq by pro-Soviet General Abdul Karim Kassem, the relationship of the Shah's regime and Iraq's succession of rulers was, for the most part, very tense and hostile. Fears of a pro-Soviet regime in Iran were nearly realized in 1958, as well, when a coup attempt led by General Qarani came within hours of succeeding. Tensions between Iraq and Iran began to ease in 1975, when the Shah and Iraqi

President Saddam Hussein settled their differences at the Conference of Algiers.[8]

The hostility of Libya's Muammar al-Ghadaffi and Palestinian Liberation Organization leader Yassir Arafat toward the Shah certainly commanded some of the Shah's attention and energy. Ghadaffi and Arafat both helped the Iranian opposition groups with financial aid and terrorist acts, to prepare them to overthrow the Shah's regime.

In addition to the threats from the Soviets, the leftist Arab nations and his enemies at home, the Shah's friends–the United States and Britain–challenged him as well. Generally the Shah's relations with the Americans and the British were friendly. Certainly no foreign ruler was ever as devoted to the United States as the Shah. The Shah's close dependence upon the United States was underscored by an almost embarrassing segment of his book[9] in which he describes himself waiting for instructions from the Carter Administration during his last days in power. The Shah himself lends credence to accusations by his critics and opponents that he was a puppet of the United States.

The Shah's dependence on the United States dates back to 1958, after the misadventure of Mossadegh, and the time of the overthrow of King Faisal of Iraq, when the Shah first turned to the U.S. for military aid. There was never any doubt of the Shah's sincerity toward the United States, although the behavior of U.S. presidents toward the Shah was often less than reciprocal. According to the Shah,[10] the relationship between Iran and the United States was closest during the administrations of Lyndon Johnson, Richard Nixon and Gerald Ford, becoming the most solid during Nixon's presidency. It was during the Republican Administrations of Nixon and Ford that the Shah reached the height of his military and economic power in Iran.

In contrast to the Republican Administration, two Democratic presidents, John F. Kennedy and Jimmy Carter, caused problems for the Shah. Kennedy disliked the Shah primarily because the Shah favored Richard Nixon in the 1960 election, and saw in Nixon a continuation of the amicable relations established during the Eisenhower administration. The Shah contributed substantial funds to Nixon's election campaign. When John Kennedy won the election, his brother Robert made serious efforts to topple the Shah in favor of the pro-Mossadegh party, the National Front.[11]

It was also during the Kennedy administration that, in 1961, a phenomenon named Ayatollah Ruhullah Khomeini appeared on Iran's political horizon, trying to initiate a revolution, albeit without any meaningful success. Shortly thereafter, Khomeini was exiled, first to Turkey, then to Iraq. It is not certain whether the Kennedy administration was directly involved in supporting Khomeini; nevertheless there is no doubt that through the intrigue of Robert F. Kennedy in Iran, the foundation was laid for Khomeini's adventure and the ultimate success of his revolution in 1979. Despite Robert Kennedy's opposition, the Shah was able to win President Kennedy's support during an official visit to Washington, which prevented further U.S. initiatives to topple the Shah.

Jimmy Carter was the second Democratic president to cause trouble for the Shah. Carter's Human Rights Doctrine contributed to the fall of the Shah. While the Shah counted Johnson, Nixon and Ford among his friends, his opinion of Carter, because of the discrepancy between his words and deeds, was lower. The Shah wrote:

> ...Carter appeared to be a smart man. My favorable impression of the American President deepened while he visited

Tehran to spend New Year's Eve with us at the Niavaran Palace. I have never heard a foreign statesman speak of me in such flattering terms as he used that evening..."Iran, because of the great leadership of the Shah, is an island of stability in one of the more troubled areas of the world," Carter said in his prepared remarks at dinner. [12]

In another section of his book, the Shah wrote of the last days of his rule:

"...After we reached an agreement with the White House, President Carter telephoned. He warmly wished me good luck and reiterated the assurances of his aides. It was the first and only time I had spoken with the president since wishing farewell on New Year's Day, 1978, when he visited Tehran." [13]

If the Shah was disappointed by Carter's attitude, other statesmen even sought to distance themselves from Carter as his presidency drew to an end. Helmut Schmidt, the former West German Chancellor, noted that, "To show yourself side by side with Jimmy Carter became a political risk for some European government leaders...for me anyway." [14] It is no secret that Carter's ineptitude was the trademark of his presidency.

The downfall of the Shah was not the only blunder to occur because of Carter and his Human Rights Doctrine. His political skills, or, rather, lack of them, contributed to the loss of a pro-U.S. Nicaragua to the pro-communist Sandinista regime. It was also interesting to note Carter's expression of disappointment upon learning that the Soviets had invaded Afghanistan. Had Ford remained president, neither Iran nor Nicaragua would have been lost, and Afghanistan would probably have been spared a decade of bloodshed.

The domination of Iran by British colonial policy began long before the reign of Mohammed Reza Shah, and continued, in complicated, shrewd and cruel ways, to be channeled through their various agents over the years until the

Shah's fall and beyond. The success of British domination of Iran for such a long period is explained by two factors: the illiteracy of the majority of Iran's population, and the shrewdness and cunning of the British policy itself.

The British imposed their policy upon Iran, particularly during the Shah's regime, through a variety of channels: Iranian politicians; clergymen; feudal landlords and anti-Shah tribal leaders; the Iranian Tudeh (Communist) Party; anti-Shah and leftist Iranian students, mainly residing abroad; and the British Broadcasting System. These were the six pillars on which the British colonial policy in Iran was built. In order to understand the extent of British influence in Iran, one must consider the significance of these six pillars.

The Shah was surrounded by a plethora of politicians, many of whom were Freemasons. These politicians severely limited the Shah's latitude in making political decisions, especially where British interests were concerned. It has been suggested that the Shah himself was a Freemason, which implies an even more direct influence on him from this mysterious organization.

On the surface, it would seem beneath the dignity of the King of Kings for the Shah to become a simple Mason, unless he could have at least held the highest position as Grand Master. Even so, he would have had to decide which of the Iranian lodges with which to affiliate: the pro-British one, or the pro-American. This would have been certain to cause conflict one way or the other. Further, it seems unlikely that the Shah would have exposed himself to use as propaganda material by the Soviets.

On the other hand, it is believed that the Shah was a Freemason who belonged to the Grand Lodge of Scotland, and that Jafar Sherif Amami, prime minister of Iran when the revolution was building, was also a member, holding a higher rank than the Shah's. To support this belief there is not a single photograph of Amami kissing the Shah's hand. This gesture was a tradition of all high officials and ministers of the government during each New Year's (Nawrus) celebration. It is only speculation as to whether or not the Shah was a Freemason. Certainly those who could confirm or deny it are unlikely to step forward.

However, there is no doubt that the Freemasons were deeply rooted in Iran, as in many other countries. The full extent of the influence of this secretive organization on the politics and economy of the world will never be known. Still, a closer look at what is known will help clarify the political environment which surrounded the Shah.

Freemasonry[15] originated in England, and branched out all over the world, except for the communist countries, where Freemasonry is supposed to have been eliminated. Interestingly, the Roman Catholic Church specifically forbids its members to join the Freemasons. With more than six million adherents representing almost every country where Freemasonry is not officially banned, the Freemasons are the largest secret society in the world.

All Freemasons have sworn, on pain of death and ghastly mutilation, not to reveal the Masonic secrets to outsiders, who are known to the brethren as the "profane". The serious nature of this cult is evident in the revenge taken upon its members who betray secrets or violate the rules of the Lodge. The penalties range from disgrace to death. However, faithful adherence to the principles and policies of Freemasonry assures a member of the support of the organization for himself and his family.

The formation of the Grand Lodge of England, the mother lodge, in London on June 24, 1717, marked the beginning of the wide dissemination of Freemasonry. The present-day fraternal order carries out its secret business and rituals in a deliberately cultivated atmosphere of mystery. Masonic Temples may be buildings specifically for these purposes, or private buildings or even hotel rooms, temporarily converted for Masonic use.

The principles of Freemasonry have traditionally been liberal and democratic. Dr. James Anderson (1684-1739), a Scottish Presbyterian minister, wrote the *Book of Constitutions* published in 1723, and revised in 1738. This book presents the by-laws of the Grand Lodge of England and prescribes religious tolerance, loyalty to local government, and political compromise as the Masonic ideal.

Masons are expected to believe in a supreme being, use a holy book appropriate to the religion of the Lodge's members, and maintain secrecy concerning the order's ceremonies.

In 1730, thirteen years after the order's origin, the first lodge was established in then-colonial America, in Philadelphia. Benjamin Franklin later joined one of the first American Lodges, and published an edition of Anderson's *Constitutions*. George Washington became a Freemason in 1752, but later refused to become the head of Masonry for the whole of the newly-formed United States of America. Freemasonry in the United States came to be organized state by state. Today, almost every state has its own Grand Lodge.

The influence of the Freemasons in the political life of the United States is widely felt. In any given year, the majority of governors, senators and congressmen are likely to be Freemasons. The United States was truly fertile ground for the brotherhood. Eight signers of the Declaration of Independence (Benjamin Franklin, John Hancock, Joseph Hewes, William Hooper, Robert Treat Payne, Richard Stockton, George Walton, and William Wipple) were known Masons, while twenty-four others, on less certain evidence, have been claimed by the brotherhood.

Seventeen U.S. presidents have been Masons, from Washington, Madison and Monroe, to Johnson, Ford and Reagan. Seventeen vice presidents, including Hubert Humphrey and Adlai Stevenson, were also brethren. One is reminded of the Freemasons every time one looks at a dollar bill: on one side is Washington's image, and on the other is the all-seeing eye, a symbol of Masonry.

In England, where the secret society of the Freemasons has its deepest roots, it has enjoyed the support and patronage of royal families since the formation of the first lodge. At present, Queen Elizabeth II is the Grand Patroness, although, as a woman, she is banned from entering a Masonic Temple.

For the first time the Grand Lodge of England may soon lack a monarchial Grand Patron. The future king of England, Prince Charles, has refused to join any secret society. Prince Philip, on the other hand, has determined not to rise in the Masonic hierarchy (the average Mason will generally proceed through the first three Degrees–Entered Apprentice, Fellow Graft, and Master Mason–while the hierarchy extends upward to the Thirty-third Degree) and thus he is inferior in rank to thousands of commoners, and cannot take the office of Grand Master. The present Grand Master of the Grand Lodge of England is the Duke of Kent, the Queen's cousin, who was installed in June, 1967.

Politicians are not the only ones who become Masons; there are Masons in all walks of life. There is also much cross-fertilization between the Masons and other groups such as the Round Table, the Lions Club, the Rotary Club, and the Chamber of Commerce.

Freemasonry is a power to be reckoned with in almost every country where it exists. For example, Francois Mitterand owes his loss in the close 1981 French presidential election to the Freemasons. Had Valery Giscard d'Estaing not become a Freemason, joining the Franklin Roosevelt Lodge in Paris that year, the election would have been won by the Socialist Mitterand.

Although the British Grand Lodges recognize over one hundred Grand Lodges, forty-nine of them in the United States, they do not have control over all of them. Most of them reflect the character and political complexion of their home countries. Far from being revolutionary, there is no organization more reactionary, more establishment-oriented, than British Freemasonry. Its members benefit from the brotherhood only so long as the status quo is maintained.

In Iran, the foreign Grand Lodges which dominated the country's political and economic scenery were the Grand Lodge of Scotland (which directly controlled the Grand Lodge of Independent Iran) and the Grand Lodges of France, Germany, and several from America. The increase of American influence in Iran began with the elimination of the feudal system through Agrarian Reform, which also destroyed one of the formerly strongest pillars of British colonial policy.

Since the Iranian Army was indirectly controlled by the United States, the Americans moved to establish pro-U.S. relationships with Iranians in important government positions. To achieve this, the Americans established Masonic Lodges in Iran, whose members included American military personnel, among them the Officers' Lodge, the Message Lodge, the Salvation Lodge, and the Square and Compass Lodge.

The pro-U.S. Masons in Iran were Iranians who had studied in the United States. It did not take long for these Iranian Masons to achieve power in Iran, gradually replacing the Iranian Masons who were under the influence of the

Grand Lodge of Scotland ̇ he pro-U.S. Masons were dollar-minded; they were more interested in making money than in political status. At meetings where deals were made among themselves, opium, alcohol and women were often available. As a result, corruption in government was widespread.

The extent of corruption in the Shah's government is found in the activities of Masons who were ministers of the government and who left Iran just before the victory of Khomeini's revolution in 1979. Prior to leaving the country, these ministers transferred huge sums of money abroad. The following figures were provided by employees of Iran's Central Bank on November 1, 1978. They show funds transferred out of Iran in just two months, September and October of 1978, to feather the nests of Iranian government ministers, many of whom were Freemasons. A certain amount of similar activity on the part of these people prior to this two-month period would not be unlikely.

In contrast to the dollar-minded pro-U.S. Iranian Masons, the pro-British Iranian Masons were older and much better financially established. An exception was Amir Abbas Hoveyda, who was more power-oriented and cared less for financial gain than did his colleagues. Although Hoveyda was prime minister for thirteen years, at the time of his execution by the Khomeini regime he possessed little worldly wealth. Nearly all of the former Iranian prime ministers (Eqbal, Ala, Alam, Amini, Mansur, and Amami) were pro-British Freemasons. Former Prime Minister Jamshid Amuzegar was a pro-U.S. Mason, and Hoveyda was pro-German. Regardless of their orientations, there is little doubt that the Iranian Freemasons not only disregarded the best interests of Iran, but also paved the way for the downfall of the Shah.

Jamshid Amuzegar, who was prime minister shortly before the revolution, stopped the salaries paid by the government to the Islamic clergy, which caused turmoil

Figure 1

Name	Minister of:	Amount Transferred, (Millions of Dollars)
Hushang Ansari	Finance	69.6
Abdolazim Valian	Agriculture	8.9
Anushiravan Puyan	Health	7.4
Manucher Shahgholi	Health	7.3
Hushang Nahawandi	Culture	7.3
Dr. Sheikhulislami	Agriculture	6.8
Dr. Jahanshah-Saleh	Agriculture	6.9
Jamshid Amuzegar	Prime Minister	5.9
Iraj Wahidi	Energy	5.0
Abdolhussein Majtabi	Agriculture	5.4
ManucherTaslimi	Commerce	5.3
Gholamreza Kianpur	Justice	5.3
Manucher Gangi	Education	5.1
Ezatollah Yazdan-Panah	Advisory	4.6
Manucher Alikhani	Commerce	4.4
Manucher Azmun	Deputy Prime Minister	4.4
Homayun Jabransari	City Planning	4.1
Ataollah Khoshrawani	Labor	3.1
Dariyush Homayun	Information	2.6
Fatullah Sotude	Post Office	2.4
Ali Asghar Hedayati	Education	2.3
Kasem Wadi	Labor	2.1
Total Transferred in two month period		$176.2 million

among the clerics and touched off the revolution.

There is no doubt that Amuzegar knew what this would lead to; he justified his actions by saying that, as a good Mason, he was just carrying out an order from a higher level. Even after all of this damage, the Shah felt that he made a big mistake in accepting Amuzegar's resignation. The Shah later wrote, "I should never have allowed this wise and unbiased counselor to withdraw."[16]

Jafar Sherif Amami, prime minister during the development of the revolution, caused friction through a series of measures he actually took against the Shah. He allowed Iranian television and the news media to criticize the Shah, thus smoothing the way for Khomeini's final victory. During this period of mounting pressure on the Shah, Amami and former Prime Minister Ali Amini (both pro-British Freemasons), along with members of the Oil Consortium, urged the Shah, in their last meeting with him, to accept a proposal of the Oil Consortium. This proposal stipulated that the Shah would not pursue greater independence for Iran's oil industry after the expiration of the current agreement, and that the Shah should renew the agreement restricting the independence of Iran's oil industry. The Shah refused, thwarting the Freemasons. All that was left to be done was to get rid of the Shah.

Dr. Mossadegh was a Freemason who earlier caused the Shah grief. His services as a Freemason in nationalization by the Iranian oil industry were well rewarded. Mossadegh was selected for this task by the British for several reasons. First, Mossadegh was old and retired, and had never before held a political position or been involved in politics. Here, he had the opportunity of a lifetime. Second, the outcome of this task did not matter to him, because he was at the end of his career. Third, the fact that Mossadegh was considered anti-British in Iran made him especially attractive as an instrument of British policy. And, finally, the British had every confidence in him; Mossadegh was a Freemason, therefore he was a trusted man. For his services, Mossadegh was promised a priceless reward: he would be made a National Hero of the Iranian People.

Evidence even exists that the Ayatollah Khomeini himself was a Freemason.[17] This raises the consideration of another pillar of British colonial power in Iran: the clergy (mullahs).

Using religion as an instrument to realize political and economic goals was nothing new for the British. They had recently discovered religion to be a handy wedge when used to inspire the conflict between the Hindus and Muslims of India. The Muslims soon gained independence from India for Pakistan. In achieving their goals, the British found the clergy very helpful, in no small part because the great majority of their religious followers were illiterate. It is not surprising that Iran became a victim of this British strategy as well. The British realized the power of the Islamic clergy in Iran, particularly that the mullahs could move masses of people in the name of *Jihad* (holy war). The British used the clergy freely, whenever they felt it necessary to do so. During the Mossadegh era, an Iranian mullah, Ayatollah Kashani, and his followers were known as British agents and served British interests to a great extent.[18]

The association of the Iranian mullahs with the British was not limited to Kashani and his followers. In general, the mullahs were considered as British agents and traitors to the national interests by the Iranian population. The mullahs received money from the British, primarily from British Petroleum. In return, the mullahs caused political havoc throughout the country upon request, through their weekly sermons. The British were quite successful in using the mullahs to enforce their policy in Iran.

The center of the pro-British mullahs in Iran was Qom, a religious city about 100 miles south of Tehran and 170 miles north of Isfahan. The mullahs and the city of Qom were so important to the British that they would resort to any means to protect the mullahs and the city from destructive influences. In the late 1950s, for instance, the Iranians tapped what was probably the largest oil reserve in the world just a few miles from Qom. The well was named Alborz. The British feared that the exploitation of this well could change the people of Qom through the influx of oil money, so they set fire to the well. After putting out the fire, the British prevented re-exploitation of the Alborz well, as well as future development in the area. Qom and its people remained pliably backward, the Alborz well has not been heard of since.

Reza Shah, the father of Mohammed Reza Shah, tried to eliminate the influence of the mullahs from Iran, and was largely successful. He warned his son against trusting the mullahs and associating with them, and urged him not to promote their cause or to support them in any way. In spite of this, Mohammed Reza Shah helped elevate the mullahs in the eyes of the Iranian people. The Shah, it seemed, was too soft on the mullahs. His prime minister, Amir Abbas Hoveyda, established a regular salary for the mullahs, in order to feed them. Further ignoring his father's advice, the Shah, upon surviving any accident (including two assassination attempts) would claim that he had been saved by the invincible hand of the Imam (son of the Prophet), which only added to the credibility of the mullahs in the eyes of the majority of the illiterate and fanatically religious people of Iran.

The feudal landlords and anti-Shah leaders of the tribes also served the interests of British colonial policy, in differing ways. The feudal landlords kept the peasants poor by exploiting them. The anti-Shah tribal leaders, many of whom were landlords as well, revolted against the central government of the Shah. The height of this resistance came during the Agrarian Reform of the Shah's White Revolution. The reason for opposition to the Shah's reform was that it would have given economic independence to the peasants and raise their standard of living, removing the control of the feudal landlords...and the British.

The Iranian Tudeh (Communist) Party always caused problems for the Shah and Iran. It was members of the Tudeh

Party who made attempts on the Shah's life. It was the Tudeh Party that sought to separate the province of Azerbaijan from Iran. There are many examples of Tudeh mischief in Iran.

One might suppose that because the Tudeh Party was a communist party, it must have been pro-Soviet. This is half true. Actually, the British helped the Soviets establish the Iranian branch of the Communist Party. It was an open secret in Iran that the members of the Tudeh Party were at least as pro-British as they were pro-Soviet. The reason for British involvement in launching a communist movement in Iran was simply to exercise control in whatever manner was available. This is nothing new for the British; they meddle wherever their interests dictate.

According to the Shah, the British thought that having agents who pretended to be anti-British would give them control of the nationalist movement, and that a hold on the Tudeh Party would give them similar leverage. The British specifically hoped to infiltrate their agents into the workers in Abadan and in the refineries and oil fields of the south.19

The Tudeh Party was mostly active during the time of Mossadegh. It surfaced again during Khomeini's revolution and had considerable power in the early days of Khomeini's reign. However, by 1985 their leaders were all imprisoned or executed by the Khomeini regime, and as a result the Tudeh Party in Iran was dissolved.

Probably the greatest source of trouble out of Iran for the Shah were the leftist Iranian students. According to the Shah, the first student demonstration against his regime broke out in San Francisco is 1962, and was financed by the major oil companies and the American CIA. The reason for the formation and financial support of the Iranian student movement was the displeasure of the oil companies and the American and British governments at the Shah's success at negotiating a 75/25 oil royalty contract (in Iran's favor) with Enrico

Mattei of the Italian oil company E.N.I. in 1957. The student movement was part of an organized and concentrated effort to discredit the Shah and his regime.[20]

Although the first Iranian student demonstration was in San Francisco, the primary office of the Confederation of Iranian Students Overseas was in London. Its founder, Richard Cottam, was a professor of political science at Pittsburgh University, with reputed connections to the C.I.A. and the British Intelligence Service.[21] Cottam did not have much difficulty forming the Confederation and recruiting students. Many of the Iranian students were less than motivated when it came to their studies; many had been away from Iran for years and still had not been accepted by a university. Both those who couldn't finish their studies and those who couldn't get started were short of money, lonely and homesick, and frustrated by the potential embarrassment of returning to Iran without degrees. These Iranian students joined the Confederation of Iranian Students Overseas hoping to overthrow the Shah and gain high level positions in the ensuing government in Iran. Some of the students realized their hopes, but the majority were totally disillusioned after the fall of the Shah and the rise of Khomeini.

Reflecting on that time, it is difficult to imagine that the British and the Americans had formed this movement and used it to their full advantage. Not only the Shah, but the Iranian students themselves, became victims of a conspiracy of the oil companies–primarily British Petroleum–British Intelligence and the C.I.A. Although the student movement overseas touched a sensitive nerve in the Shah, his regime never tried to discover the reasons for the students' dissatisfaction. On the contrary, the regime fought back with repressive measures, driving the students deeper into the web of their sponsors.

The British Broadcasting Company (BBC) has always served as a long arm of British colonial policy in many countries. It was no different in Iran. The BBC would broadcast daily in the Persian language, beaming commentaries, news items and other programs to stir up political unrest in Iran against certain individuals or the government in general, against whomever had displeased the British government. These underhanded broadcasts were quite successful in arousing the Iranian public whether with fact or rumor. Not surprisingly, most radios in Iran were tuned to the BBC. The repressive policies of the Shah's regime left it ill-equipped to counter BBC propaganda in Iran.

Much has been written about the Shah's ambitious plans which were intended to make Iran the Great Civilization, via industrialization, creation of a strong infrastructure, and education for all. These ambitions were praiseworthy, and the Shah's efforts definitely made great strides toward transforming a backward and underdeveloped country into a progressive, modern nation, especially during the 1970s, when oil money began to flow into Iran as never before.

As laudable as his ideas for the Great Civilization were, the Shah's attempts to realize them were often shortsighted and doomed to failure. For example, consider his campaign to build nuclear power plants throughout Iran, four of which were under construction and fourteen in the planning stage at the end of his reign. The folly of the Shah to invest billions of dollars to supply the energy demands of a country rich in fossil fuels speaks for itself, especially at a time when the industrialized countries were seeking to curtail further development of nuclear power plants.

Like many other leaders, the Shah often relied on unrealistic and deceptive figures and statistics. He and his ministers were always proud to talk of economic growth and the increased standard of living in Iran. While much was achieved, even more could have been accomplished. The fact remains that there was a tremendous discrepancy between the statistics and reality. For instance, the improvements in the standard of living in Iran are attractive on paper, when statistical averages are considered. In fact, eighty per cent of the wealth in Iran was distributed among twenty per cent of the population. Naturally, the eighty per cent of the population who held twenty per cent of the wealth were less than satisfied. And, while most of the cited improvements favored the wealthier twenty per cent disproportionately, the Shah's reform policies, including profit sharing for the workers, served to turn the more fortunate twenty per cent of the population against him as well.

The Shah's detachment from everyday reality would cause him to look upon a model factory or village as exemplary of the situation of the whole country. If a few workers owned cars or hired domestic help, the Shah saw this as evidence of national prosperity. He wrote proudly, "When I visited a sugar factory near Quchan in 1973, I learned that eighty per cent of the workers owned automobiles and that fifty per cent of them hired domestic help."[22] He seemed to overlook the fact that the domestic help's wages must have been menial at best, and that their standard of living could not have been equal with the national norm; worse, he imagined that the circumstances of the workers in Quchan were representative of those throughout the country.

The principles embodied in the Shah's White Revolution would have provided a firm foundation for Iran's becoming a modern and progressive country, if these principles had been successfully implemented. Although it is undeniable that much was accomplished, the overall record of shortfall and failure of the White Revolution was furthered by lack of support and by the corruption of the people in charge.[23]

Corruption was rampant in Iran. It was impossible to accomplish anything without bribery, whether a government minister or down to the lowliest government employee. This corruption also spread to members of the Shah's family. None of this helped the Shah at all; rather it fostered animosity, frustration and resentment among the people of Iran.

The Shah was often accused of excessive generosity, or outright squandering, in his allocation of the natural, and monetary, wealth of his nation. And, in fact, practically every country in the world enjoyed some benefit from the wealth of Iran. The West received huge orders for arms, technology and large projects such as nuclear power plants, most of which did little for the people of Iran, but helped to keep the economies of the West afloat.

The Russians also shared in the wealth of Iran, by getting much- needed natural gas at below-market prices.

The British fared the best of all foreign governments; they managed to finance the exploration for, and exploitation of, their North Sea oil with generous financial help from Iran. After the success of the North Sea venture, the British neither reimbursed Iran, nor gave Iran any share of the bounty. To this day, there is controversy over whether the Shah's give-away policies were somehow intended to benefit Iran, or himself.

The SAVAK (from the initials in Persian for the Organization for State Security and Information) represented the darkest and most brutal side of the Shah's regime. It is true that after the adventures of the Mossadegh era the Shah needed to consolidate a degree of political power through intelligence gathering, in order to prevent further strife in Iran, from sources internal or external. The SAVAK's intense repression of the people, however, far outweighed its largely ineffective actions against the subversive activities of the communists.

The Shah claimed that at the end of 1978, the SAVAK employed less than 4,000 persons.[24] The Shah's account may be true, because at that time his regime was near collapse, and since the activities of the SAVAK were in decline, it is possible that its payroll was, too.

However, in 1976, William Butler, chairman of the International Commission of Jurists' Executive Committee, claimed that the SAVAK had up to 200,000 full-time employees "operating in every nook and cranny of the Iranian system, and also in many places, especially where Iranian students [were] congregating, both in the United States and in West Germany, and in other parts of the world."[25] Since the leaders of the Iranian Army were under the SAVAK's control, it could be said that another 400,000 people were operating, at least indirectly, for the SAVAK. Brutal intervention of the military in the political sphere of Iran became a pervasive characteristic of the Shah's rule. Whenever necessary, the military and the SAVAK were used to crush and demoralize the opposition of whatever political stripe, to manipulate the behavior of the citizenry, and to control and redirect public opinion in the name of the regime's stability and security. In these endeavors, the military and the secret police imprisoned, exiled, or executed hundreds of people, almost indiscriminately.

The SAVAK was, unlike any other institution in Iran, a well-structured and highly effective organization, which spread its tentacles not only throughout Iran, but across the world as well. If the SAVAK did not have the tight control of events abroad that it enjoyed in Iran, at least very little happened without the SAVAK's advance awareness.

Two aspects of the SAVAK's operations were especially controversial and disgusting to the people of Iran: the

SAVAK's shrewd cultivation of the politics of distrust, so that the people came to believe that they were being constantly watched by the SAVAK's ubiquitous secret members; and the SAVAK's pursuit and persecution of the students as well as opposition political groups. In furthering the latter aims abroad, the SAVAK used Iranian embassies and consulates as outposts.[26]

Geneva served as the European headquarters of the SAVAK, and was the preferred venue for SAVAK conferences and seminars.[27] The SAVAK's activities in Iran and abroad were open secrets. On June 1, 1976, thirteen Iranian students pushed their way into the Iranian Consulate at Avenue Campel 24, Geneva, and gained access to secret SAVAK documents. This was the first time that the public had had hard evidence of the SAVAK's activities and methods abroad, and of its secret codes. The code names for the Shah were *Mansur* (victor) and *Niknam* (a person of good reputation). The code for SAVAK branches was *Lubiya* (beans), for SAVAK agents *Tamispanje* (clean finger), the SAVAK's European headquarters in Geneva *Simin* (the golden) and Tehran headquarters *Lahut* (divinity).

The SAVAK was first established in 1953. It started to grow rapidly once it gained the approval of the Iranian Parliament in 1957. The American C.I.A. not only helped to set up the SAVAK, but also directed its operations and provided all the specialized equipment the SAVAK needed to fulfill its tasks. With the devotion of C.I.A. Director Allan Dulles, the SAVAK quickly became one of the most feared secret services in the world. The Israeli Secret Service (MOSSAD) also helped with the training of SAVAK agents, who became the MOSSAD's best sources of information on Iran and the surrounding Arab countries.

The SAVAK's notoriety grew, as did that of its leaders. The

SAVAK became famous for treating those opposed to the Shah and his family very brutally. According to Amnesty International, torture was commonplace, with the menu including electric shock, anal penetration with broken bottles, weights suspended from testicles, and rape. The rape of their victims was an art so refined that SAVAK agents used trained animals such as bears for this purpose. Women would be raped before their husbands' eyes.[28]

The SAVAK's first chief was General Teimour Bakhtiar. The leader of Iran's most feared organization became the most feared man in the country. Bakhtiar held his position until 1962, when his ambitions to challenge the Shah and possibly replace him became obvious, which posed a serious threat to the monarch. Bakhtiar was sent into exile in Iraq, where several years later he was assassinated by agents of the organization he had helped to create.

The second chief of the SAVAK was General Hassan Pakravan, who later served as Iran's ambassador to Paris. Pakravan, who was trained in the United States, was an expert in psychological warfare. He was responsible for introducing these methods to the SAVAK, which used them most effectively.

General Nematollah Nassiri's friendship with the Shah helped him rise to become the next SAVAK chief. Nassiri was notoriously gruesome, sadistic, and corrupt. The amount of money he alone transferred out of Iran in 1978 during the months of September and October was in excess of $55 million. One can only speculate as to how much he may have salted away prior to that time. Like many other generals, Nassiri was executed when Khomeini came into power.

In June, 1978, when the revolution was gaining momentum, Nassiri was replaced by General Moghadam, about whom little is known. He was executed like his two predecessors, and the SAVAK was dissolved by the Khomeini government.

There are no known statistics concerning the SAVAK's victims. Amnesty International estimated that the number of people tortured was between 25,000 and 100,000.[29] According to the Shah, the number of people arrested for political reasons between 1968 and 1977 was exactly 3,164.[30] Assuming that this number is correct, it does not account for the periods of 1957 to 1968, or 1977 to 1978, especially active times for the SAVAK.

As much as the end of SAVAK was welcome news to the people of Iran, it was a demise in name only. A new secret police organization, SAVAMA, replaced the SAVAK almost overnight. SAVAMA utilized almost all of the SAVAK's former employees.

The SAVAK had become so pervasive in Iranian society that the majority of Iranians did not trust one another, let alone the government. This cost the Shah much support from the general population. Things did not change under SAVAMA.

The function of General Hussein Fardust in this connection proves interesting. Fardust was a close friend of the Shah from the time they were children six years of age, and attended LeRosey, an elite school in Switzerland. Very little has been written of Fardust, except that he was one of the Shah's most trusted advisors. Fardust betrayed the Shah in the end and became the head of Khomeini's SAVAMA.

General Fardust rarely showed himself in public, and his picture almost never appeared in any publication. Whenever his name was mentioned in the media, it carried an air of mystery. The secret service career of General Fardust actually began with his involvement in the SAVAK as an assistant to General Bakhtiar. Later, he became General Pakravan's deputy. In the SAVAK under Nassiri, Fardust refused to remain a deputy because he held a higher military rank than Nassiri. As a result, he left the SAVAK. Later, Fardust was appointed Director of the Imperial Inspection Commission. The Commission's function was to study the programs of the ministries, monitor their operations, and correct serious organizational problems. In effect, Fardust was the Shah's watchdog over the entire government. Fardust remained in charge of the Imperial Inspection Commission until shortly before Khomeini's takeover, when he changed sides to become the first chief of SAVAMA.

After nine years under the Islamic Republic of Iran, Fardust died as mysteriously as he lived. No one knows the reason for or manner of his death. All that is known is that a year before he died, he donated all of his land and wealth to the needy people of Iran–probably, it is speculated, against his will.

One wonders what possible justification Fardust could have had for the betrayal of his friend, the Shah, for whom it must have been especially painful. Recent evidence suggests that Fardust may have been involved in the attempt on the Shah's life on April 10, 1965, when a soldier broke into the Marble Palace and opened fire with a machine gun. However, Fardust's involvement was never proven.[31]

A different side to the story of General Fardust holds that Fardust was a man of vision and integrity, who tried very hard to remedy the grievances in Iran.[32] According to this story, Fardust warned the Shah in 1973 that Iran would be lost within five years if things did not improve and the people remained dissatisfied. At about the same time, Fardust reportedly advised the Shah that the government should try to discover the reasons for the students' opposition to the regime, rather than simply to fight them with the SAVAK.

Furthermore, it is reported that Fardust and his Imperial

Inspection Commission prepared mountainous files detailing the abuses and corruption of the regime's officials and presented them to the Shah. The files disappeared, or escaped attention, as soon as the Shah forwarded them to Hoveyda, the prime minister. The dissatisfaction of the Iranian people continued to grow.

The greater part of the corruption in the Shah's regime continued primarily because of three people: Asadollah Alam, a former prime minister, then court minister; Houshang Ansari, the finance and economic minister; and Sharif Amami, prime minister and president of the Senate. The corruption in the government was by no means limited to these three; with a few exceptions, almost every minister and official in the government was involved as well: Amuzegar, Khosrowshahi, Agheli, Setude, Mahdavi...the list seems endless. These are but a few of the names that Fardust presented to the Shah.

As to the accuracy of these charges, there is no doubt; the evidence is strong. The question remains, why in the end did Fardust betray the Shah? Without trying to justify or defend his actions, one can speculate that Fardust changed his position because of disillusionment and desperation with the Shah's ineffectiveness on the one hand, and the corruption of an unimprovable regime on the other. If General Fardust was a man of such principles and integrity, it is unlikely that he found himself any happier in the Khomeini regime, except perhaps that he was alive.

Another series of factors worked to the Shah's disadvantage and were decisive in bringing about the collapse of his regime. The duality of the Shah himself was one such factor. The Shah presented himself as a modern monarch, trying to industrialize his country, surrounding himself with educated people, and bringing modern technology into Iran.

At the same time, he told the people on many occasions that his life had been saved in numerous crises, accidents and assassination attempts, by Imam Reza or Imam Abbas, two descendants of the Prophet Mohammed.[33] In doing so, he apparently hoped that the Iranian people would be convinced that he was protected by divine power. There is no doubt that the Shah faced many life-threatening situations in his life, but by stressing divine intervention as his salvation, the Shah only helped to elevate the credibility of religion–and the mullahs–in the minds of ordinary Iranians. When it came time to learn whether the Shah or Khomeini was closer to divine power, Khomeini emerged the winner, and the Shah had helped to dig his own grave.

The imperial family itself was another source of problems for the Shah, and increasingly detracted from his image in the country. For example, according to Fereydoun Hoveyda, brother of the former prime minister Abbas Hoveyda: "One of the Shah's brothers, Mahmoud Reza, had been given permission to cultivate the poppy and to sell opium. He used to sell a large amount of opium on the black market and at high prices. As a result, a scandal broke out in Switzerland in 1972,when one of the members of the Shah's retinue, Amir Hushang Davalou, with a warrant out against him for drug dealing, got away from the local police. The Shah himself drove him to a plane waiting at the Zurich airport, under the eyes of the helpless police."[34]

The outrage of the Iranian people at countless similar examples was so extensive that finally, in mid-1978, the Shah signed a decree regarding "The Ethical Conduct of the Imperial Family."[35] However, this gesture was too little...too late. The people were geared to oust the Pahlavi Dynasty from Iran.

The frustration of the Iranian people with the Shah's family was not caused so much by the imperial family's involvement in the international drug trade; this had little effect on the daily lives of the people. The royal family's brutal and corrupt business practices and other activities in Iran affected the people more directly. Although almost all the members of the imperial family were involved in corrupt business dealings, the Shah's twin sister Ashraf, with her son, Shahram, and one of the Shah's brothers, Gholam Reza, had extremely low reputations and were the most hated by the populace.

The Shah himself was also a target of harsh criticism, for exploiting the wealth of the nation for his own interests. According to a list published shortly after the revolution,[36] the Shah's wealth included, in part, four banks in Iran, major shares in New York banks and other banks in Iran and abroad, shares in eleven large companies in Iran, majority stockholdings in sixteen factories, ownership of four transport ships ranging from 33,000 to 53,000 tons, two famous restaurants, twelve renowned hotels, six night clubs and gambling casinos, ten palaces,[37] and six universities.[38]

The Shah also owned properties, buildings and landholdings overseas. His famous Colorado residence, located on a large piece of land near Aspen was valued at over a million dollars, and he also owned several large villas in the state of Florida. The Shah possessed over four million acres of land in Australia, which was used for raising lamb for consumption in Iran. In addition, the Shah maintained foreign bank accounts totaling five to seven billion dollars, from which substantial amounts were diverted to investment projects in the United States and other western countries.

Empress Farah, the Shah's wife, although trying to help him, inadvertently caused the Shah additional difficulties and weakened his position even further. Before becoming the Queen of Iran, Farah was a simple Parisian art student with only two years of university studies to her credit. In Iran, she felt her knowledge and appreciation of art enabled her to promote art and artists in her country. She wanted to intro-duce European and modern art to Iran, but in so doing forgot that Iran was not France. She lost sight of the fact that Iran is an Islamic country whose population is fanatically religious.

In the summer of 1976, Empress Farah organized the Festival of the Arts in Shiraz. This festival included pieces of erotic acting on stages that were built in the streets of Shiraz, so these scenes were exposed to the public. Farah was proud of her accomplishment, but the next day the mullahs in Shiraz organized mass prayer in repudiation of her and her arts. In the course of Khomeini's revolution, mass prayer and demonstrations were two tools which were used successfully to move the people. The revolution began in the city of Shiraz.

Another well-intentioned act of Empress Farah which backfired occurred in the early days of the revolution. Her telephone call to Ayatollah Shariatmadari, one of Iran's most powerful mullahs, to plead for assistance in preventing the progress of the revolution, damaged the Shah's position and gave a big boost to the mullahs'. In this call, she introduced herself as *Madere Reza* (mother of Reza, the crown prince) rather than as *Shahbanoo* (wife of the Shah). To appreciate the significance of this, and to realize the extent to which the Empress was debasing herself and literally crawling to the mullahs, it is necessary to understand a social custom of the Islamic Iranians, particularly of the lower classes. Among these people, a woman is seldom called by her own name. When women are introduced, or introduce themselves, they are referred to as the mother or sister of a male in the family, reflecting the Muslims' belief that women are inferior to men. Even if this custom were practiced by the Empress, which was unlikely for a woman of her class, simply referring to herself as the Shah's wife would have been polite enough in talking with the lower class mullah. Calling as the mother of Reza, then, must have been as humiliating for the *Shahbanoo* as it was detrimental to the Shah.

To save her husband's throne, Farah used any means, including whatever she could do to please the mullahs. In 1978, Farah visited Ayatollah Abul-Qaem Mussavi-Khoi, the Grand Ayatollah of Najaf, a holy city for the Shi'ite Muslims, and asked for his support at that critical time. As a token of her appreciation, Farah offered that her husband's regime would remain passive, and even provide tacit assistance, in the elimination of the followers of the Baha'i faith in Iran, if this would please the mullahs. Although there is no definitive proof of Farah's involvement, shortly after this visit thirteen Baha'i houses in Shiraz were burned to the ground and seven Baha'is were killed by a mob.

Using the Baha'is as scapegoats to appease the mullahs was nothing new; prior to this, it had happened several times. Most notably, in 1955, a large scale attack on the Baha'is was led by a mullah named Mohammed-Taqi Falsafi, with the Shah's blessings and approval.

Falsafi, who was Khomeini's representative in Tehran, began his nationwide campaign against the Baha'is by using Radio Tehran. He called upon the Iranian people, in terms none too subtle, to kill the Baha'is and destroy their holy places. Promptly, many Baha'is met that fate and their holy places were destroyed. The bodies of Baha'is were disinterred from their cemeteries and mutilated. Shops, houses and farms were plundered. Their crops were burned and livestock destroyed. Young Baha'i women were abducted and forced to marry Muslims. When matters seemed to be getting out of hand, the Shah relented and stopped the mobs.

The Shah's tools of repression included not only the SAVAK, but a political party he formed on March 4, 1974. True to form, the Resurgence Party did nothing to strengthen the Shah's position. Conversely it prompted people to label him insane. Prior to 1974, politics in Iran were pretty well rigged. The Iran-e Novin Party always represented the majority (the government's point of view), and the Mardom Party offered token opposition from the minority. In the new Resurgence Party (Hezbe Rastakhiz), the Shah merged these two parties into one, and banned all other parties. He called upon all Iranians to join this party and to participate in politics, either through the party's "Right" wing, or its "Left" wing. He branded those who opposed his new party and the government as non-patriotic, and asked them to either cease their political activities or to leave Iran. He warned that if they remained in Iran, they would suffer penalties for their wrongdoing.

The Shah was seen as obsessed with building up his military forces, which he ultimately raised to the sixth greatest military power in the world. The Shah's military might even made the Americans and the Soviets nervous. It is true that after the British withdrawal from the Persian Gulf, the U.S. decision to make Iran the policeman of the region had a great deal to do with the Shah's efforts to fill his arsenals with the world's most sophisticated weaponry.

Between 1971 and 1978, the Shah spent $19 billion for the purchase of weapons from just the United States. Certainly this amount of money would have gone a long way toward fulfilling the Shah's dreams of the Great Civilization. And, was the military build-up was necessary? The expenditures on arms drained the economy's resources and replaced them with economic problems. Iran's military escalation, of course, fostered distrust on the part of her neighbors. After all, Iran acquired more military might than all of her neighbors' combined, with the exception of the Soviets. This served to heighten tensions in the region, for which the arms companies had ready relief.

Although his critics[39] never missed an opportunity to point out the consequences of the Shah's obsession with modern weaponry, they had little to say about the Shah's reasons for beginning the massive build-up. The Shah was caught in the confluence of three powerful forces: the arms companies, the oil companies, and the United States government.

The arms companies were booming as never before, and they wanted even more business. Once it was decided that there were big profits to be made in the Middle East, the Shah was doomed. If he didn't buy his quota of arms, and become the leading military power in the region with an awesome deterrent force, then he would have to buy to defend himself against his Arab neighbors. It was clear that the arms companies were going to sell plenty of weapons in the Middle East, one way or the other.

The oil companies were looking for ways to make more money, and the best way to get a lot of it was to raise prices.

The United States wanted a force for stability in this oil-rich and strategically valuable area between the Soviets and Africa, the Soviets and the oil fields of the Middle East, and between the Soviets and the Indian Ocean. On May 30, 1972, President Richard M. Nixon and Secretary of State Henry Kissinger landed in Tehran fresh from the Moscow Summit. In their meetings with the Shah, he was given a green light to raise oil prices significantly to underwrite his purchases of arms, which made everybody happy. Kissinger crowed:

"The vacuum left by the British withdrawal, now menaced by Soviet intrusion and radical momentum, would be filled by a local power friendly to us. Iraq would be discouraged from adventures against the Emirates in the lower Gulf, and against Jordan and Saudi Arabia. A strong Iran could help damp India's temptations to conclude its conquest of Pakistan. And all of this was achievable without any American resources, since the Shah was willing to pay for the equipment out of his oil revenues.[40]"

The arms companies made a lot of money. The oil companies made a lot of money. The Shah got a lot of weapons. The U.S. got a policeman in the Middle East. The citizens of the Western industrialized nations, and thus the entire world, paid for it, as did the Iranians, who did not benefit from the increased oil income.

There seems to be no limit to the creativity and persistence of the arms market. The recent Iran-Contra scandal of the Reagan administration in the United States, where weapons were sold to Iran for profits to buy more weapons for distribution to the Nicaraguan "Contra" forces, was far less devious than the everyday dealings of America's peace-loving allies. Despite laws specifically forbidding sales of arms to countries at war, Swedish, Spanish, French, and Austrian companies wove intricate webs of deceptive transactions through layers of intermediaries worldwide to supply both sides in the Iran-Iraq war. The famous pacifist, late Prime Minister Olof Palme of Sweden, was one of the world's most successful arms salesmen,[41] and his peace-loving country was a leading beneficiary of the Iran-Iraq war. Where money and jobs are concerned, principles command very little attention.

Interestingly, the Swedish National Police placed an advertisement in the *Herald Tribune* (Paris) as well as in an Iranian newspaper, offering a reward of $8,200,000 from the Swedish government for information leading to the solution of the murder of Palme in Stockholm on February 28, 1986.[42]

The Shah could only be force-fed weapons and prevent war in the Middle East for so long. Once the flow of income to the arms industry started to slow down, or at least such a trend was foreseeable as it became evident that the Shah would be ousted, a different tactic to recycle the region's oil revenues through the arms industry was adopted. All the arms dealers had to do was to make Iraq's president, Saddam Hussein, cocky enough to attack Iran, which he did.

As a result of the war, all the prior stocks of weapons were destroyed, and more weapons sales started to roll in. Every country involved benefitted, except Iran and Iraq.

Although the Shah's arms purchases were ruinous to the economy of Iran, the Shah's power as a result of them prevented a war in the Gulf region during his reign. Once he was gone, the ensuing war brought destruction to Iran and Iraq for nearly a decade.

In the end, the Shah's arms build-up was of even less use to him than it was to his country. The Iranian Army, one of the most powerful in the world, and a showcase of modern military technology, was impotent in the face of an illiterate mob of religious zealots with vintage light arms.

If the Shah's controversial and hypocritical nature did not always engender animosity toward him, it certainly did not win him any friends. On the one hand, the Shah authorized the establishment of casinos and pleasure resorts (several of which he owned), the most notable being a gambling establishment on the island of Kish in the Persian Gulf. This was a vacation center for the rich sheiks of the Gulf region, for whom Air France's Concorde flights were used to import women chosen by the famous Madame Claude of Paris.[43] On the other hand, the Shah professed to be a very religious, faithful and devoted Muslim, and claimed to always carry a copy of the *Holy Koran* on his person.[44]

It is difficult to imagine that the pleasure palaces of Iran did anything to enhance the Shah's image in Iran; Islam specifically forbids alcohol and gambling.

Another example of the Shah's hypocrisy was the manner in which he treated his aides. There were many instances that proved that, when it was a question of survival, past services rendered and old friendships hardly mattered. Perhaps the achievements of Amir Abbas Hoveyda, whose lifetime of loyal service to the Shah included thirteen years as prime minister, were little to be proud of. But his imprisonment as a scapegoat in 1978 was a total disgrace, and showed the basest side of the Shah's character. Certainly, Hoveyda was responsible for his contributions to Iran's miseries, but so were others in the forefront, including the Shah himself. The fact that Hoveyda was the most devoted puppet the Shah ever had might even be considered to ameliorate Hoveyda's guilt somewhat.

Amir Abbas Hoveyda was killed on April 7, 1979, by a burst of machine gun fire after a short mockery of a trial in one of Khomeini's so-called Islamic Courts.[45] No one knows what would have become of Hoveyda had the Shah's regime survived. It is probable that he would have had a spectacular trial, been rehabilitated, and emerged as a national hero. The consequences of the Shah's betrayal of Hoveyda were even more severely devastating for the Shah himself. The Shah's treatment of his loyal prime minister prompted the departure from Iran of a large number of high-ranking government officials who had no desire to share Hoveyda's fate. The Shah's isolation deepened.

Hoveyda was an interesting man. He was known as an opportunist, obsessed with maintaining his position as prime minister. He tried very hard to keep everyone satisfied. For example, in May, 1975, at a party Hoveyda gave in honor of Iranian students studying abroad, one of the students asked to see the progress achieved in Iran. Hoveyda replied, "Gentlemen, from tomorrow you are my guests." The next morning a private plane was made available to fly this group all over Iran. Wherever they went, the treatment was royal. There is no doubt that this special attention was designed to keep them happy, and that had it not been for the demonstrations of the Iranian leftist students abroad, they could not have dreamed of such courtesy from the government.

Not only did Hoveyda tolerate the corruption of his cabinet ministers, and that of the whole regime, he even established a substantial fund from which the mullahs were paid monthly salaries. This was later eliminated by Amuzegar. Hoveyda's efforts to please the mullahs even went so far that he built a mosque in the prime ministry, and made prayer mandatory for his employees. He also created a new position of an advisor on religious matters and, of course, that position was awarded to a mullah. In order to appease the Islamic clergy and to obligate them to him, Hoveyda spent large sums repairing old religious buildings, and erecting new ones.

Hoveyda's devotion to the mullahs was so strong that he decided to remain in jail during the transition from the Shah's government to Khomeini's when he could easily have fled from prison. Hoveyda's brother wrote: "The wardens of the prison which held my brother and some of the other people arrested by the Shah disappeared as if by magic, and some of the prisoners escaped. My brother stayed - somebody had to show a little courage, after all. He had already refused to allow himself to be broken out of jail by an action group organized by some friends and relations. He made a telephone call to Khomeini's headquarters and decided to give himself up to the new authorities in hope that there would be a public trial where he would finally be given the chance to explain himself.[46]"

Hoveyda had every reason to expect preferential treatment at the hands of the mullahs; they owed him a lot. He later discovered that the mullahs' gratitude for past services was even more dangerous to one's health than was the Shah's.

Despite his denials, Hoveyda was often accused of being

a Baha'i. In reply to this, he declared, "I am a believer [Muslim] and have gone to Mecca."[47] In addition, Hoveyda was related maternally to Ayatollah Khoi. Even without consideration of his extensive favors to the mullahs, their labeling Hoveyda as a Baha'i was baseless. The suspicions that Hoveyda was a Baha'i stemmed from the fact that his father was a Baha'i who was excommunicated from the Baha'i Faith for his involvement in politics, which is specifically prohibited by the tenets of the Baha'i Faith. Hoveyda's father never practiced the Faith thereafter, though there are indications that he sympathized with Baha'ism. However, this did not necessarily make Hoveyda a Baha'i, and his career would have caused him excommunication from the Faith in any event.

The Shah was not only surrounded by corruption, but by flatterers, dissemblers, opportunists, and others who would do anything to please the Shah and the royal family as long as they could derive material benefit for themselves. These people filled the ranks from prime ministers, cabinet ministers and generals, down through ambassadors, consuls and their deputies, and so on throughout the regime. The lower-ranking among them rarely had the opportunity to see the Shah personally, but they never missed a chance to demonstrate their subordination to him.

As a result, the Shah was virtually imprisoned in a world which was not real. The illusory world of the Shah expanded in direct proportion to the growing, endless list of titles bestowed on him by his fawning subjects. Glorious as the titles heaped upon the King of Kings were, one of the better ones was the last one, *Arya-Mehr*–the Love of the Aryans. Blinded by pragmatic loyalty which he imagined was true, the Shah did not recognize that his minions lacked integrity. These people had a way of distancing themselves from their positions and turning to treachery as soon as their benefits were jeopardized.

General Fardust's switch of loyalty to the Khomeini camp is a prime example. And there were two of the Shah's personal pilots, who flew the dethroned monarch and his family to Egypt on January 16, 1979. After arriving in Egypt, the pilots took off in the Shah's jet and defected to Iran, hoping to join Khomeini and enjoy similar positions, or better, in the new regime.[48]

Using his authority, General Ghara-Baghi prevented military action against Khomeini. Ghara-Baghi was the main reason that one of the world's most powerful armies was defeated without defending itself at all. The Shah called him a traitor.[49] Unlike the Shah's other generals, Ghara-Baghi's life was spared and he was allowed to leave Iran in safety. His savior was Mehdi Bazargan, Khomeini's first prime minister.

The about-face of government officials during this transition from the Shah's regime to Khomeini's was so base that it borders on the disgusting. For example, the Shah's last consul general in Hamburg, West Germany, always had the Shah's portrait prominently displayed in a gold frame on his desk. While the Shah was in power, the Consul spoke highly and respectfully of the monarch. But on the very day that the Shah left Iran, the Consul replaced the Shah's portrait with that of Shahpur Bakhtiar, the last prime minister under the Shah. With the change of portraits came a change of consciousness. Immediately the Shah was a bloodthirsty man who had looted the country, and Bakhtiar was going to cast off the miseries created by the Shah and be the savior of Iran. Just a few weeks later, when Khomeini came into power, his portrait replaced Bakhtiar's in the golden frame, bringing with it a change of rhetoric: "Thanks be to God, that the Shah and his adherents are gone. Now the Master (Khomeini) will bring peace and stability to Iran. Please God, may my life be a sacrifice unto him!"

Khomeini, at least, was smart enough not to trust these traitors and deserters of the Shah. As long as he needed them, he used them. When they were no longer useful, he discarded them or sent them to the gallows. One of the famous turncoats, General Fardust, established SAVAMA for Khomeini, and died mysteriously after being forced to donate all his wealth to the needy people of Iran. The Shah's personal pilots, who had deserted the Shah in Egypt, were executed shortly after their return to Iran. The consul general in West Germany was discharged from his post, and presently runs a small grocery store in Canada. The consul, and others like him, could at least have salvaged some scrap of dignity by remaining loyal to the Shah to the end. Because they were part of the Shah's regime, they should not have expected any leniency from Khomeini's.

Of course, those turncoats had too long been part of the privileged twenty per cent of Iran's population to be completely realistic. It could not be expected that they would be strong supporters for the Shah in the face of Khomeini's adventure. Khomeini, after all, had the support of the eighty per cent of the population who had to make do with twenty per cent of the wealth. Today in Iran, twenty per cent of the population still controls eighty per cent of the wealth. The difference is that today's privileged class is the former bottom twenty per cent of the people of the Shah's Iran.

In retrospect, a whole series of factors was instrumental in bringing about the collapse of the Shah's regime. Escalating economic problems, the Shah's repressive policies, and the corruption of the entire regime caused a widening chasm between the Shah and his subjects. Also, the British government and British Petroleum were preparing a grand scheme to topple the Shah and his government. Attacked from so many sides at once, the Shah could hardly hope to keep his throne.

The foundations for the Shah's removal were laid at a conference of the major oil companies in London in November, 1975. The conference was hosted by British Petroleum and the British government itself. The oil companies met to decide on, and coordinate, policies in response to statements made by the Shah. In one recent speech, the Shah, addressing the major oil companies, demanded that the oil companies leave existing oil resources to their countries of origin, and that they should search for new sources of oil.

This demand by the Shah signaled a great threat to the interests of the major oil companies, especially to those of British Petroleum in the Persian Gulf region. Of equal concern to the British were Iran's demands to participate not only in the production of North Sea oil, in which project Iran had invested heavily, but as a partner in the refining of it as well.

The final decisions of the conference were to try to discourage the Shah from his course through the involvement of the Iranian Tudeh Party and the Confederation of Iranian Students Overseas in creating uproar in Iran and in discrediting the Shah overseas. It was further decided that if these efforts did not soften the Shah's stand, then the mullahs should get involved, and ultimately, if necessary, the Shah should be replaced.

The activities of the Confederation of Iranian Students Overseas began in earnest in 1977. Their actions against the Shah were supported by British Petroleum, as well as by other oil companies and the American C.I.A. They were guided by Richard Cottam, former U.S. Attorney General Ramsey Clark,[50] a member of the British Foreign Ministry, a representative of British Petroleum, as well as Abrahim Yazdi, who was one of Khomeini's three top aides. These were concerted efforts to discredit the Shah throughout the world. Demonstration after demonstration was arranged.

Amnesty International also got into the act. Countless bulletins were issued about the SAVAK and the brutality of the Shah's regime. Human rights activists demanded to visit the prisoners in Iran. Yet, it is interesting to note, Amnesty International and the human rights activists quieted down after the revolution, while the world continues to witness a regime whose cruelty and brutality make the Shah's pale by comparison.

It is undeniable that these actions touched the rawest nerves of the Shah's regime. A further blow was dealt by President Jimmy Carter and his Human Rights doctrine. Inside Iran, the Tudeh Party and the Mujahidin -e-Khalq (Fighters for the People) began a series of terrorist acts.

The Shah's last days began on January 4, 1979, when the heads of state of the four major Western powers flew to Guadeloupe for a summit at the invitation of French President Valery Giscard d'Estaing. At this meeting, British Prime Minister Callahan convinced President Carter and Giscard d'Estaing to approve the ouster of the Shah. West German Chancellor Helmut Schmidt was basically against such a plan; he did not see Khomeini as a viable alternative to the Shah, nor did he want to jeopardize West Germany's profitable business relationship with Iran. The French president was less worried about such matters; he was expecting preferential treatment of France by Iran under Khomeini. France had been Khomeini's host during his stay in Paris, before he could return to Iran as a leader of the revolution. Since Carter, Callahan and Giscard d'Estaing had already set their minds on toppling the Shah, Schmidt had little choice but to agree.[51]

At the same time as the Summit of Guadeloupe, Carter dispatched Air Force General Robert E. Huyser, then Deputy Commander of the U.S. forces in Stuttgart, West Germany, to Tehran. He arrived on January 4, 1979, and remained for a month. Huyser was sent to contact the Shah's top military officers, unite them in support of Prime Minister Bakhtiar's government, and to encourage the military in Iran to transfer its loyalties from the Shah to Bakhtiar. Huyser's mission was not successful. On the contrary, he literally contributed to the disintegration of the Iranian armed forces and to the arrests of many high-ranking military leaders.[52]

In his book,[53] General Ghara-Baghi accuses General Huyser literally of bringing about the downfall of the Shah and bringing Khomeini to power. Since Ghara-Baghi was involved in the daily affairs of Iran, he must have been somehow confronted with Huyser's undertakings. Ghara-Baghi states that Huyser was unqualified for such an important task, which ultimately led to the failure of his mission and the loss of Iran altogether. He claim's that Huyser's description of his mission in Tehran, as described by Huyser[54] does not always reflect the actual events and that Huyser twisted the truth to his advantage.

One question which has never been answered satisfactorily, is why the Shah, with the backing of his powerful military forces, simply yielded to his opponents. Many explanations have been offered, none entirely convincing. One scenario has that the final push on the Shah came in the form of a threat from the United States. In late 1978 and early 1979, Reza Pahlavi, the Shah's eldest son and the crown prince of Iran, was in the United States. The Carter Administration, as the story goes, let the Shah know plainly that if he delayed his departure from Iran, his son would have a mysterious accident leading to his demise. With the crown prince as hostage in the United States, the Shah decided to leave Iran.

The Shah and the Shahbanoo left Iran for the last time on January 18, 1979. That had to be the worst year of his life. Harassment followed him wherever he went. The world's treatment of a former head of state was sad to behold, especially from his friends and allies, notably the United States. Mexico, Panama, and the Bahamas outdid even the United State; they were simply after the Shah's money and tried to take advantage of his situation.

The Shah must have been astonished to see how quickly his former friends turned against him as soon as circumstances warranted. It must have been especially disappointing to see Jordan's King Hussein distancing himself. President Carter was another example of human callousness.

Henry Kissinger wrote: "America has little to be proud of in our reaction to his [the Shah's] overthrow...History is written by the victors, and the Shah is not much in vogue today. Yet it hardly enhances our reputation for steadfastness to hear the chorus today against a leader whom eight presidents of both parties proclaimed–rightly–a friend of our country and a pillar of stability in a turbulent and vital region.[55]"

After all the humiliation the Shah went through, his truest friend turned out to be Anwar Sadat. The president of Egypt hosted the Shah for the remaining days of his life. To add to his humiliation, the Shah was gravely ill, suffering from cancer.

Mohammed Reza Shah, the Shahinshah (King of Kings), finally died in Egypt in 1980, and was buried in Cairo. His death ended the Pahlavi Dynasty, and one of the most tragic lives of recent history. What stunned the world the most–and particularly the Iranian intellectuals–was not that the Shah fell, but that he fell to an old lunatic cleric named Khomeini.

4

KHOMEINI AND THE ISLAMIC REPUBLIC OF IRAN:

The Results of Ignorance

*T*he West, and the Iranian people, really didn't know what they were getting when Ayatollah Ruhullah Mussawi Khomeini came to power. For a glimpse at the intellect of the man whose Islamic Republic replaced the Shah's regime, consider the following excerpts from one of his most famous books, *Tuzih-al-Massael (The Interpretation of Problems)*. In the chapter on women with whom marriage is forbidden, Khomeini teaches:

Problem #2394: If a person, before marrying his cousin, has intercourse with her mother (in this case his aunt), he cannot marry the cousin.

Problem #2395: If a person marries his cousin and before having intercourse with her he first sleeps with her mother, even so the marriage is valid.

Problem #2396: If a person has intercourse with a woman other than his own aunt, it is advisable not to marry her

daughter; but if he marries the woman and then has intercourse with her mother, the marriage will remain valid. The
marriage can also remain valid if a person after the wedding
first sleeps with the bride's mother and subsequently with his
wife. In this situation, however, it is advisable to divorce the
wife.[56]

In the more than six hundred book pages Khomeini offers
similar lessons for every aspect of human life, from marriage,
divorce, eating, drinking, business, praying, cutting one's
nails, going to the toilet, menstruation, sex during fasting,
including sex with animals, and beyond. Altogether the
book deals with 599 essential subjects and their related problem areas.

Although *The Interpretation of Problems* has all the in-
gredients of a compendium of dirty jokes, it actually lacks
the basic guidelines for behavior which it was written to
provide. As soon as this book hit the marketplace in 1979 it
became the butt of jokes among the Iranians and foreigners
as well. After a short time, the book became an embarrassment to the Islamic Republic of Iran, and was withdrawn
from circulation. It is practically unavailable today, and the
rare copies that do surface command big prices. Khomeini's
book was translated and published in many countries.[57]
Even *Playboy* magazine published excerpts in a *Playboy
Report*.[58]

Predictably, in the beginning, no one took Khomeini
seriously. Everyone was amused by the old man who, armed
with his speeches and prayers, sought to topple one of the
mightiest regimes in the world. Later, during the occupation
of the U.S. embassy in Tehran, and the dramatic hostage
crisis, the world realized that he was deadly serious.

Although some of Khomeini's ideas are behind the times,
and some of his teachings may be quaint and amusing, the

fact is that he ruled not only Iran, but almost all of the fanatical adherents of Islam, with a power seldom seen before; certainly greater than that of any previous Iranian ruler. The reason for this is his ability to control the minds and hearts of the millions of fanatically religious, but illiterate, people of Iran. During the revolution, this power won over the Iranian intellectuals and so-called progressive groups as well.

During the Shah's era, the women of Iran enjoyed almost all the freedom and privileges of European and American women. During the revolution, as a gesture of support for Khomeini and of protest against the Shah, many Iranian women started to wear the *chador* (veil). In their revolutionary euphoria, they did not consider the future. Once the revolution succeeded, the Khomeini regime commanded all women to wear the chador, and lowered their status to a point where they are servants and objects of pleasure for men. Khomeini also abolished the Shah's restriction that men must be monogamous; under Khomeini and Shi'ite Muslim teachings, men may have up to four permanent wives, and unlimited *sighe* wives (temporary contract wives). The *sighe* marriages are temporary arrangements, usually for money, whereby a man marries a woman for a specified number of days or months. Since the revolution, Iranian women have not been allowed to enjoy the simple personal freedoms of dress, make-up, public or family life, or career, that was taken for granted under the Shah. But Khomeini's grip on every facet of human life goes far beyond the suppression of women.

Ruhullah was born in 1902 in Khomein, a typical small religious Iranian city between Isfahan and Qom. He was only one year old when his father, Seyyed Mustafa, was killed in a dispute over a piece of land. This later led to a persistent myth that Reza Shah Pahlavi, the late Shah's father, had killed Khomeini's father, and that this was why the son, Ruhullah, was the sworn enemy of the Pahlavi regime. Chronology belies this; Seyyed Mustafa was killed in 1903, and Reza Pahlavi became Shah in 1922, nineteen years later. Neither was aware of the other's existence. This myth was almost certainly created by his followers.

However, in the period between 1927 and 1940, when Reza Shah brutally put down a revolt by the mullahs, Reza Shah first became aware of Ruhullah Khomeini.

Khomeini began his theological studies in the city of Arak, near Qom, in 1918. After a year at the seminary, he followed his master, Sheik Abdul-Karim, to Qom, and settled there. It was not until 1942, with the publication of his first book, *Kashfe Asrar (The Unveiling of Secrets)*, that Khomeini's radical nature gradually became apparent.

Until 1960, Khomeini's radicalism did not create much of a problem for Mohammed Reza Shah. In 1960, Khomeini's second book, *The Interpretation of Problems*, was published. It was also in 1960 that the Shah introduced his land reform program, which generated the mullahs' resistance to the Shah under the leadership of Khomeini.

The first serious collision of Khomeini and the Shah happened in 1961. A year after this clash with the Shah, Khomeini received the title *Ayatollah*, which means "the symbol of God" and allows a mullah to decide about theological matters. In 1963, Khomeini's campaign against the Shah's reforms, especially those relating to land reform, the emancipation of women, and the employment of non-Muslims–in particular Baha'is–in government positions, reached its peak. Khomeini was arrested and thrown in jail.

Khomeini's imprisonment caused an uproar throughout

the country, and many people were killed. The people's demands for his freedom led to Khomeini's release to house arrest in Tehran. When events returned to normal, Khomeini moved back to Qom and once again started to criticize the government. This time his criticism concentrated on the law granting diplomatic immunity to American military personnel. The Shah's patience ran out. Khomeini was arrested again, and this time he was exiled to Turkey.

While in exile, Khomeini tried to find a city he could use as a base for his religious and political activities. Back in Iran, his followers were stirring up more trouble, including the assassination of Prime Minister Hassan-Ali Mansur, at which time Amir Abbas Hoveyda became prime minister. Meanwhile, the controversy between Khomeini and the Turkish government resulted in his obtaining an exit permit, enabling him to finally settle in the holy city of Nadjaf, Iraq, in 1966.

For a while things were quiet with Khomeini. As time went by the number of people who attended his seminars and listened to his sermons grew in size as well as in radicalism.

The next attempt to shake the Shah's regime from Khomeini's exile in Iraq occurred during the years of 1969 to 1971, when he criticized the Shah for wasting the country's wealth on the celebration of the 2,500th anniversary of the Persian Kingdom. This did not create much of a problem for the Shah, since it was also at this time that oil revenues were beginning to increase.

For the next few years, from 1972 through 1976, when the Shah and Iran were at the height of their euphoria of wealth and power, Khomeini was conducting more and more seminars in Nadjaf, which were attended by more and more anti-Shah students.

Also during this time, Khomeini published the first edition of his book *Valayate Faqih (Domination of Theology)*. Later, a much more polemic version of this work was published under the title of *Hukumate Islami (Islamic Government)*. By now, Khomeini's goals were already defined, but very few people paid attention to his words.

One of those few who did was Abrahim Yazdi, a naturalized Iranian. Yazdi was married to an American woman, was living in the United States, and had known connections to the CIA. In 1977, Yazdi encouraged Khomeini to intensify his campaign against the Shah. He claimed that the new American government, under Jimmy Carter, wanted to oust the Shah. As a reaction to Yazdi's encouragement, Khomeini wrote a piece in which he declared the Shah dethroned and removed from his responsibilities. This occurred just as the Shah and Carter were visiting each other.

Had it not been for an article insulting to Khomeini which appeared in Tehran's daily newspaper *Ettelaat* in 1978, history would probably have been written differently. The publication of this article caused a storm of upheaval which spread from the city of Qom to Tabriz, and later to most other Iranian cities. Hundreds of people lost their lives in confrontations with government forces. Khomeini demanded the removal of the Shah, but instead Khomeini was expelled from Iraq. He tried to take refuge in Kuwait, but was denied entry. He ended up in France, where he continued his attacks on the Shah, now with the collaboration of the French media.

As Khomeini's campaign against the Shah gained furor every day, the Shah and his officials were, more and more frequently, making the wrong decisions at the wrong times at the wrong places. According to General Ghara-Baghi,[59] nothing was normal in the government of Iran at that time. The Shah was indecisive. A few of the ministers were not cooperative, or failed to perform their duties. Some of them tried to bring about the collapse of the government so that they themselves could become prime minister. The Shah left the country in early 1979, after turning over power to Shahpour Bakhtiar, who had been his opponent for twenty-five years.

With regard to Bakhtiar's short term as prime minister, it must be noted that had he not made the Shah's departure a condition of his accepting the premiership, but had insisted that the Shah remain in Iran, it is likely that the country would have been spared the destruction and madness Khomeini wreaked upon it. However great the Shah's passivity and indecision were at the time, his presence in Iran would have been substantial. Further, the Shah's presence would have provided the military's support for Bakhtiar, and prevented the military itself from falling into disarray. In fact, Iran's military was lost when the Shah departed; its leader was gone.

Although Bakhtiar portrayed himself as one of the most honest and patriotic of the country's politicians, he became the victim of his own misjudgments. Bakhtiar, who had fought for his ideals for twenty-five years, failed in his chance of being an effective leader. His brief tenure as prime minister was disastrous.

Iranian sources[60] state that Bakhtiar's animosity toward the Shah and the Pahlavi Dynasty was much stronger than his interest in saving the country. It is said that Bakhtiar's primary goal in accepting the premiership was to use the position to fulfill his lifelong dream to oust the Shah. According to these same sources, Bakhtiar asked Khomeini whether his contribution to ousting the Shah had pleased Khomeini, and what Khomeini wished him to do next. Bakhtiar is portrayed as a British agent fulfilling a master plan. It is not difficult to accept this scenario in the context of what later transpired.

Two weeks after the Shah left Iran, Khomeini returned in triumph to Iran, and was welcomed by millions of his countrymen. The cabinet of the newly-installed Bakhtiar government fell apart and the victory of Khomeini's revolution was complete. Shortly thereafter, Khomeini, through a plebiscite, dissolved the monarchy and instituted the Islamic Republic of Iran.

The formation of the Islamic Republic brought with it disappointment and disillusionment. The dark and disastrous record of Khomeini's regime is endless, including massive and barbaric executions of people from all walks of life, among them: the Shah's generals and officials of his government; Baha'is[61]; Mujahidin-e-Khalqs[62]; and even Khomeini's former top aides, notably Sadegh Ghotbzadeh. Other dark deeds included taking of American hostages, acts of terrorism abroad, creation of economic chaos, corruption, and the destruction of the economic and social order of Iran. Khomeini led Iran into isolation from the world community, and oversaw the deaths of hundreds of thousands of youths in the bloody war with Iraq.

The disappointment of those who had followed Khomeini with full heartfelt support must have been deep. The predicament of the multitude that simply followed the Khomeini tide is understandable. What is perplexing and inexcusable is the behavior of those who should have known better, and who, worst of all, later tried to justify their behavior, to express their disappointment at having been deceived, or to rise up in opposition to Khomeini. The list is numerous; the most famous among them are:

- **Hassan Nasih,** a lawyer and politician who was among the first group of politicians who visited Khomeini in Nauphle-le-Chateau, France, and declared his support of him. Before long, he joined the opposition to Khomeini;

- **Admiral Ahmad Madani,** Khomeini's defense minister, who ran against Bani-Sadr for the presidency of Iran. Known for being anti-Shah, and by the Shah's generals as being mentally disturbed, Madani now lives in France;

- **Mehdi Bazargan,** who, as Khomeini's first prime minister, was a powerless puppet, now living in Tehran;

- **Sadegh Ghotbzadeh,** who, along with Yazdi and Bani-Sadr, was among Khomeini's closest aides. Ghotbzadeh served as foreign minister for a short time, was accused of plotting a coup against Khomeini, and was subsequently executed;

- **Karim Sanjabi,** one of the leaders of Mossadegh's National Front Party and foreign minister under Khomeini, who had visited Khomeini in France and pledged his support. After a brief term in office, he joined the opposition; and

- **Abolhassan Bani-Sadr,** who became the first president of the newly-formed Islamic Republic of Iran, and was one of Khomeini's three closest aides.

The relationship between Khomeini and Bani-Sadr was one of father and son. This relationship probably played a significant role in the outcome of the election in which Bani-Sadr became the first president of the Islamic Republic. Bani-Sadr's presidency was colorless; he was overpowered by the mullahs, who occupied all the important government positions. Once it appeared that Bani-Sadr was getting too close to the army, and suspicions arose that he was plotting a coup, Bani-Sadr chose to flee rather than to risk his life.

The history of Bani-Sadr is not unique. Quite often a few people will support a leader, launch a successful coup, and almost immediately begin to establish their former leader, now the new ruler. The outcomes of these power struggles vary widely. However, lucky is the opposition leader who escapes alive to continue his struggle from abroad, writing books about the revolution.

In Bani-Sadr's book *Khinyanat be Omid (Betrayal of Hope)*[63], he attempts to justify his involvement in Khomeini's revolution. The title represents his great disappointment in his master. Nevertheless, one finds it difficult to believe his words, in reviewing his actions.

There is no doubt that Bani-Sadr masterminded the revolution, with the collaboration of Yazdi and Ghotbzadeh, and that of the three he was Khomeini's favorite. Khomeini was a spiritual father to Bani-Sadr, and the picture of Bani-Sadr bending to kiss Khomeini's ring while Bani-Sadr was president of Iran remains fresh in the memories of Iranians. In his book, Bani-Sadr portrays himself as a liberal, who had high hopes for Iran and placed importance on the role of women in society.

When Bani-Sadr proclaims himself a liberal, it is at best a wishful claim; it was impossible to be a liberal and support the Islamic revolution. Khomeini's concept of Islam was ultra-conservative, fascist, and, worst of all, inhumane. Bani-Sadr, like most of Khomeini's former supporters who later changed their minds, tried to blame his collaboration with Khomeini on his own ignorance of Khomeini's real intentions, or Khomeini's change of face after his revolution succeeded. Such explanations are simply unacceptable, since Khomeini had published his entire program in various books, all in great detail. It is impossible to believe that these people, and Bani-Sadr in particular, were unaware of Khomeini's plans.

Consider these words of the self-proclaimed liberal Bani-Sadr: "Whatever Imam Khomeini should say, I will accept it" and "For me to be an Iranian without Islam makes no sense."

Bani-Sadr, in his capacity as chairman of the Revolutionary Committee, was literally responsible for the appointment of Sadegh Chalchali, Khomeini's famous hanging judge, to carry out the massive executions during the early years of the regime.[64] Bani-Sadr, due to his support of Khomeini, shares responsibility for all of the post-revolutionary miseries of Iran.

Dr. Abrahim Yazdi was another dubious figure of Khomeini's Islamic Revolution. Yazdi was discovered in 1960 by Richard Cottam during his visit to Iran. In the years that followed, Cottam began to groom Yazdi, along with three other Iranians, Ghotbzadeh, Bani-Sadr, and Amir Entezam, for future tasks.[65] (Entezam later became labor minister in Bazargan's government, was unmasked as a CIA agent, and imprisoned for life.) Cottam's efforts began to pay dividends in 1975, when he and Yazdi, Entezam, Ghotbzadeh and Ramsey Clark went to Nadjaf to visit Khomeini. Their purpose was to persuade Khomeini to lead a revolution against the Shah. The British, Americans and the major oil companies led by British Petroleum, were interested in getting rid of the Shah, or at least pressuring him to follow policies more to their liking.

Khomeini did not require much persuading.

Yazdi's first public activity against the Shah took place during the Shah's visit to Washington, D.C., in November, 1977, accompanied by Empress Farah. Yazdi was a naturalized citizen of the United States, and worked at a pharmaceutical research facility in Houston, Texas. He was also president of the Organization of Islamic Students in North America. Yazdi arranged an anti-Shah demonstration in front of the White House while the Shah was visiting Jimmy Carter. When the scenes of this demonstration were aired on

Iranian television, which showed the Shah protecting his eyes from the effects of tear gas used to disperse the demonstrators, the Iranian people could not believe their eyes. The Shah's vulnerability was dramatically demonstrated.

Yazdi contributed to the revolution's success not only as its master planner, in collaboration with Bani-Sadr and Ghotbzadeh, but as a source of endless encouragement to Khomeini. Yazdi's enthusiasm was a significant contribution to the cause.

After the revolution, in one of Khomeini's Islamic Courts, Yazdi put harsh questions to General Rahimi and General Nassiri, condemning their crimes against the Iranian people in their service to the Shah. In doing so he portrayed himself to the media as an Iranian folk hero. Regardless of the fact that the new regime's atrocities made the previous one's look less harsh, Nassiri and Rahimi were executed.

The greatest disservice Yazdi performed for Iran was his creation of the most cruel and barbaric arm of the Khomeini regime, the Islamic Revolutionary Guard. The Revolutionary Guard conducted all the arrests and executions after the revolution, and continued to spread repression throughout the country. To assure that the Revolutionary Guard would possess the level of zeal and brutality he deemed appropriate, Yazdi filled its ranks with the dregs of society: former prostitutes, brawlers and thugs of every description. One of the greatest successes of the revolution was its cure for the inferiority complexes of the members of the Revolutionary Guard.

For his contributions to the revolution, and the stunning success of his Islamic Revolutionary Guard, Yazdi was awarded by being appointed foreign minister, upon the resignation of Karim Sanjabi. Yazdi is currently in the United States, foddering for whatever his next mission may be.

There was no shortage of dubious characters in Khomeini's regime. One prominent member was Mohammed-Hossein Beheshti, not only an Ayatollah himself, but, until his death in June, 1981, the second-most powerful man in Khomeini's government. Beheshti was a devoted student of Khomeini. He taught English in the schools of Qom, and later received a well-paid position in the Ministry of Education as the "Religious Advisor". Subsequently, he became the *Imam Jomeh* (religious leader of a community) of Hamburg, West Germany. While he was in Hamburg, Beheshti served as an agent of the SAVAK, reporting on the anti-Shah elements in Hamburg.

It has been reported[66] that the judgment for the assassination of Prime Minister Hassan-Ali Mansur was given in a secret Islamic Court under the chairmanship of Beheshti and another Ayatollah, Motahari, and was approved by Khomeini himself. The involvement of Beheshti and Motahari in Mansur's assassination remained a well-protected secret, until the Islamic Republic was formed; the assassins would never reveal who masterminded the deed although they were interrogated and tortured before they themselves were executed. After the revolution, the secret was heavily publicized, to gain credit for the action.

Between his duties as a Khomeini supporter and Ayatollah, and his work for the SAVAK, Beheshti found time to transfer huge amounts of money, Iranian rugs, and valuable antiques from Iran to European countries. The port of Hamburg became a floodgate for the rugs and antiques.

In June, 1981, the Mujahidin placed a bomb in the Iranian Parliament in Tehran. Beheshti and seventy other prominent leaders were killed in the blast.

The terrible phenomenon of the Khomeini regime was a stocky and bloodthirsty Ayatollah who served as Khomeini's

hanging judge: Sadegh Chalchali. During the early days of
the revolution, Khomeini needed a judge to expedite the
dispensation of Islamic justice. This process was required
due to the vast number of enemies of the revolution, potential
threats to the revolution, and people whose elimination
would please the radicals and leftists. The task demanded a
ruthlessly efficient administrator, who had a taste for blood.
Of the many mullahs to whom this responsible position was
offered, Chalchali was the most qualified.

It did not take long for Khomeini to discover just how
efficient Chalchali was. Justice hopped along in his kangaroo
courts; Chalchali sentenced the victims to death faster than
they could be executed. In one thirty-minute stint in Sanadaj,
on October 1,1979, Chalchali sentenced fifty-three prisoners
to death. With this industrious pace of thirty-four seconds
per adjudication, Chalchali could never have been accused
of denying the defendants a speedy trial.

Chalchali was responsible for executing seventy of the
Shah's generals. Chalchali was also responsible for the execution of Prime Minister Amir Abbas Hoveyda, as well as
countless thousands of others who were, in many cases,
completely innocent.

During his tenure, Chalchali's brutality and inhumanity
were boundless; he was obviously sadistic, and enjoyed his
work. The enthusiasm he had for his task was not surprising
considering Chalchali's background and the constant encouragement coming from Khomeini.

The class of people from which the mullahs in general, and
Chalchali in particular, come is the basest of all classes in
Iranian society. Had these people not managed to become
mullahs, and begun to climb in the Islamic hierarchy, their
place in Iranian society would have made them servants,
errand boys, or some other sort of menial laborers. The

position of mullah in the lower strata of Iranian society had a certain amount of petty power, and attracted ranks of thugs, charlatans, and ne'er-do-wells, who reveled in brutality and cruelty.

With such a labor force available to him, it is understandable that Khomeini called upon the mullahs for their services to the new Islamic Republic. The epitome of these mullahs were Chalchali and the mullahs of the Islamic Courts. Their mandate was broad: to eliminate all the *Musfed fil arz* (the depraved of the earth). Chalchali interpreted it broadly enough to include hanging children of all ages, especially twelve- and thirteen-year-olds. Chalchali and his master, Khomeini, were not the least bit concerned with world opinion, as long as their butchery guaranteed the elimination of anyone who posed the slightest threat to their own existence. Khomeini and Chalchali demonstrated their originality and ingenuity by initiating the hanging of children–the deterrent effect of which did not fail to grasp the people's attention.

Chalchali remained Khomeini's favorite henchman until 1982, when it was discovered that his industriousness had also enriched his personal wealth to thirty million dollars from bribes and ransoms. The ultimate proof of Khomeini's love for Chalchali is that Chalchali lost his job, but not his life, liberty or property.

The corruption and moral depravity of the mullahs were open secrets, and they became vicious when Khomeini came to power. None who were unfortunate enough to fall under their ruthlessness would ever forget the experience; many did not live long enough.

The mullahs base character was shaped by profound desire for money, power, and insatiable lust for women. Perhaps not all mullahs were driven to enrich themselves and indulge their carnal desires; certainly there were a few individual exceptions. But, the mullahs had the same kind of destructive effect to Iranian society as that of cancer cells in the human body.

Although the clergy of most any of the world's religions strive for material gain and power, the mullahs raised these practices to new heights of achievement. It is common practice for the clergy to use the faith of the believers, and the name of God, to demand contributions; one only needs to consider the cathedrals, mosques and other properties owned by religious institutions, and where the money came from for the acquisition.

In Iran, the Islamic religion was much the same before Khomeini. However, with the success of the revolution, the mullahs added political power, and the authority to use armed force, to their religious power. Consequently they gained tremendous economic power as well. No longer limited to the charitable contributions of their poor flocks, or their modest government salaries which had been cut off anyway, the mullahs undertook to exercise their power to enrich themselves in a most efficient manner. They began by assuming control of the black market which thrived in the aftermath of the revolution. The mullahs had control over the Islamic Revolutionary Guard. The Islamic Revolutionary Guard had authority over everyone else, and exercised this authority proudly and zealously. The Islamic Revolutionary Guards could enter a house at random, and take whatever they wanted without fear of repercussion, if they shared the loot with the mullahs.

Khomeini's Shi'ite Islam affords plenty of opportunity for men to enjoy the pleasures of the flesh without overstepping the bounds of sin, and the mullahs have always taken advantage of such broad opportunities to lead exemplary lives.

Islamic marriage laws not only allow a man to have four permanent wives, but as many temporary ones as he wants–or can afford–as well. Marriages for a specific period of time, usually for a specified price, are known as *sighe* marriages. A practical aspect of this form of matrimony is that, upon expiration of the contract, it is illegal to keep the woman any longer.

Are *sighe* marriages hypocritical or merely a legalized form of prostitution? This would overlook the tremendous amount of sin which is avoided by sanctifying otherwise venal behavior, and the tremendous convenience they provide to Islamic men as a means of channeling their desires into expressions of their faith. The mullahs, with their special positions of power and their responsibility to live exemplary lives, are at the forefront in the practice of marriage, to an extent that lesser Muslims can only envy. *Sighe* marriages do not always involve previously married women. Often, teenage girls are helpless victims, virtually sold into marriage.

In 1979, at the height of the victory of the revolution, Ayatollah Mahmud Saduqi of Yazd, who was very close to Khomeini, scandalized the city by marrying a fourteen-year-old girl, when he was in his sixties. He had noticed the beautiful Zoroastrian girl, had a strong desire for a religious experience, and ordered his aides to bring her to him. Ayatollah Saduqi, himself the father of four daughters, married by *sighe* the unfortunate girl, who had been orphaned and was living with her grandmother. The marriage was for one month, and took place without the consent of the girl or her grandmother. After raping and abusing the helpless girl, he scrupulously held to the term of the contract and divorced her right on time without consideration. The hopes of a girl for the future was destroyed to satisfy the basest needs of a vicious and predatory man; at least it wasn't sinful.

Ayatollah Saduqi was so hated by the people of Yazd that a young Mujahid was willing to sacrifice his life to eliminate Saduqi. At a religious meeting, the young Mujahid jumped up with a live grenade in his hand, embraced Saduqi, and blew himself and the Ayatollah to bits. The young Mujahid became a posthumous hero of the people of Yazd, with a place deep in their hearts. The elimination of Saduqi, however, had no salutary effect on the behavior of the other mullahs. Admittedly, the case of Saduqi may have been extreme, but it was not an isolated one.

The animosity of Khomeini and his followers toward mankind in general is well evidenced, but their hatred of the Baha'is in particular is deep-rooted. As early as the spring of 1955, a mullah named Ali-Akbar Halabi visited Khomeini to discuss a very important plan. His ambitious goal was to eliminate all of the Baha'is in Iran, unless they would be willing to renounce their faith and become Muslims. So began the systematic harassment and elimination of the Baha'is.

After winning Khomeini's support, Halabi discussed the plan with General Timur Bakhtiar, the chief of the SAVAK, and with General Batmanghelitch, the minister of foreign affairs. It was not too difficult to gain the support of the two generals when Halabi promised to have the mullahs support the Shah's regime against the leftists in the country. As soon as the agreement was sealed, the killing and harassment of the Baha'is began in earnest, under the leadership of another mullah, Mohammed-Tagi Falsafi.

Baha'is are considered to be peace-loving people. Since their faith prohibits participation in politics, they are seldom the targets of political persecution. However, in Iran the Baha'is are considered heretics by the mullahs, and therefore, in the mullahs' opinion, deserving of the death penalty.

The widespread hatred of the Baha'is in Iran grew out of years of negative propaganda from the mullahs, which did not fail to affect the population. The Baha'is were targeted by the mullahs for two reasons. First, the Baha'is hold that there is no need for clergy in religion; anyone can establish a relationship with God without having to resort to a clergyman, and that there is no need for a middleman between God and mankind; secondly the underlying implication that clerics were corrupt and immoral.

This certainly enraged the mullahs. The Baha'is believe that if the Baha'i religion promoted the clergy, the mullahs would have embraced Baha'ism themselves, and promoted it throughout Iran. This is unlikely, however, because of the second cause of the mullahs' hostility toward the Baha'is: one of the main principles of the Baha'i Faith is "equality between men and women". This was impossible for the mullahs to comprehend since they considered women only to be objects for male pleasure.

A broader, underlying cause of the mullahs' hatred of the Baha'is was that the Baha'is enjoyed a much higher socioeconomic status than the mullahs through higher education, hard work, and trustworthiness. They gained an elite status of sorts, especially in the eyes of the mullahs, who came from the lowest class in Iran.

Fortunately, Ali-Akbar Halabi's efforts to annihilate the Baha'is in Iran did not succeed, and in the early years of his campaign he was handed a great defeat by one of his closest accomplices, a mullah named Seyyed Abbas Alawi. Alawi was just as determined as Halabi to exterminate the Baha'is completely, but at the height of his hatred of the Baha'is, Alawi came in contact with Baha'i literature, and with the Baha'is themselves. It was not long after that Alawi became a Baha'i himself. Not only did he convert from Islam, he wrote an exemplary book entitled *Bayan-e-Haghihat (Proclamation of the Truth)*, in which he vigorously criticized the Islam of the mullahs. Seyyed Abbas Alawi became one of the most forceful advocates of the Baha'i Faith, and propagated the religion until he died.

Since Iraq's President Saddam Hussein started the devastating war between Iraq and Iran, no one can completely blame Khomeini for this catastrophe. Nevertheless, Khomeini's intransigence in continuing the war unless Hussein was eliminated, contributed to the continuation of this war from 1980 until a tenuous cease-fire was established in mid-1988. Over half a million men and boys were killed, with scores more maimed and crippled. Ironically, when Khomeini first returned to Iran after the revolution, he visited Tehran's largest and best-known cemetery, Behesht-e-Zahra. In his speech on that occasion, Khomeini condemned the Shah for having buried the victims of his regime there. Perhaps he felt that space would be needed for his own victims. Behesht-e-Zahra was soon filled to capacity.

The fact that Iraq dared to attack Iran at all does not say much of Iraq's esteem for Khomeini. During the Shah's rule, Iraq would never have attacked; they only dreamed that they might possibly defend themselves in the event of an Iranian invasion. When the Shah was in power, Iran's army could have overrun Iraq in twenty-four hours. Although the Shah was accused of being ambitious and having expansionistic ideas, he did try to avoid conflict with Iraq for at least two reasons: to prevent massive death on both sides, and to prevent the destruction of the Abadan oil refinery. Once Khomeini stepped in, both of these nightmares became realities.

Iraqi President Hussein overestimated his country's capabilities, and underestimated Iran's. Hussein was certain that a short military attack on Iran at the height of the political confusion would result in Iran conceding to Iraq not only the Shatt-al-Arab, but also part of the province of Khuzestan, which Iraq considered part of the Arab nation. Another reason Hussein decided to attack Iran was the direct encouragement of the United States. Obviously, President Carter hoped that the war would help bring about the release of the fifty-two American hostages held in Tehran.

Hussein began the war on September 22, 1980, and was convinced that it would be finished by the fifth of October. Needless to say, his optimism was unfounded; the war continued for nearly eight years. Although Hussein had hoped to take advantage of the chaos in Iran, the war actually helped to consolidate Khomeini's position. With the army occupied fighting Iraq, the chance for a military coup against Khomeini was obviated. With the people of Iran polarized against the Iraqis the economic disorder was understandable, and Khomeini escaped blame.

Although Iraq's invasion of Iran brought advantages to Khomeini's regime, the Iranians paid a terrible price in what turned out to be their bloodiest war in a century. Khomeini's relentless continuation of the war cost the nation its wealth and decimated its most precious asset: its youth. Khomeini's power to move masses of people to march forward to be slaughtered was without peer. He motivated them by the promise of admission to heaven, and gave each soldier a little key to wear around his neck. All the keys to heaven carried the imprint, "Made in Taiwan".

Khomeini and the mullahs owed one explanation to the people: if death on the battlefield leads directly to heaven, and if heaven is the greatest desire of every Muslim, then why shouldn't Khomeini and the mullahs have availed themselves of this golden opportunity and saved the country countless misery and destruction?

But the clever mullahs made the people believe that dying in the name of God, and for His cause, was a blessing reserved only for those worthy of it, and that not everyone was deserving of such great honor. To glorify their immolation, the victims' names were decorated with the word "Martyr". Today there is hardly a single family in Iran that does not have at least one "martyr". Other nations take pride in the achievements of their youth in science, sports and the arts; in Iran, the national pride views the martyr-count as evidence that its youth were devoted and faithful Muslims.

One of the most outstanding characteristics of Khomeini's Islamic Republic of Iran is the utter absence of even the most basic human freedoms. The women must wear dresses covering the entire body, and veils covering the face. Make-up and perfume are forbidden. Men are not allowed to wear short-sleeve shirts. The clothes of men and women may not be dry-cleaned together. Listening to music is taboo. The list is endless. Should a person violate any of the rules of conduct, members of the Islamic Revolutionary Guard, who are charged with enforcing their observance, quickly take appropriate action against the hapless miscreant. The punishment can vary from a simple reminder (but usually involves physical pain and humiliation), a fine, or even imprisonment.

Under the rules of Islamic Iran, every woman must carry a small "book of conduct" with her at all times, in which all her wrongdoings are registered. The "book of conduct" must be presented upon demand, for example, to a prospective employer.

Although the Islamic Republic's strict code of conduct may be viewed as evidence of its promotion of a high standard of morality, the truth is that its purpose is to keep the people so preoccupied with superficial details that they have no time to worry about important matters, such as the absurdity of their situation. If the repression of personal freedom were lifted in Iran today, the mullahs would be gone by tomorrow.

Economic conditions under the Islamic Republic are no better than the state of personal freedom. During the reign of the Shah Iran was, if not perfect, the envy of much of the world. The economy prospered, providing full employment for the Iranians, and for tens of thousands of "guest workers" from other countries. With Khomeini's revolution came high rates of unemployment and inflation; shortages and absences of goods and medicines; domination of the black market over all primary goods such as bread, milk and meat; and the drastic devaluation of the currency. If the Shah's officials exploited the wealth of the country for their own purposes, Khomeini's gang bested them. Where the Shah's economy provided full employment in building the nation, Khomeini's offered unlimited opportunities for military service.

The political balance sheet of the Khomeini regime after a decade of power provides a confusing picture of disarray. Consider the developments since the revolution began to heat up in 1978: Khomeini's engagement to lead the revolution was associated with the promise that, once the revolution succeeded in toppling the Shah from his throne, Khomeini and his fellow mullahs would return to the city of Qom; the mullahs would stay out of politics, and the political stage would be occupied by professional politicians.

In the beginning, Khomeini kept his promise to some extent, allowing Mehdi Bazargan to become the first prime minister of the new Islamic Republic of Iran. But after only a few years, in the face of escalating threats to the mullahs, Khomeini decided to get rid of the politicians altogether, and to replace them with mullahs. His former promise of noninterference in politics forgotten, Khomeini set out to consolidate the mullahs' position in the political arena in a characteristically radical manner: all of his former officials, who had supported him so vigorously, were just as vigorously thrown out of their positions, and many were executed. Among those who lost their positions to the mullahs were: Bazargan, the first prime minister, who was removed; Bani-Sadr, the first president, who fled Iran to escape the gallows; Sanjabi, the first foreign minister, who was removed; and Ghotbzadeh, the second foreign minister, who was executed.

In order to prevent further challenges to his authority and to stem the growing tide of threat, Khomeini executed or imprisoned the leaders of the two dominant political groups in Iran, the Iranian Tudeh (Communist) Party, and the Mujahidin-e-Khalq (Fighters for the People). According to a political publication of the Mujahidin, which was documented with names and photos of the victims, in this brief period Khomeini's Islamic Revolutionary Guards executed over 12,000 of the Mujahidin. Masud Rajavi, the leader of the Mujahidin, was more fortunate. He escaped Iran with Bani-Sadr.

The politics of Khomeini's Islamic Republic did nothing to win prestige or friends for Iran abroad. The United States, which hoped to block the expansion of the Soviet Union through the creation of the Islamic Republic as a sort of Islamic greenbelt, received the first dividends from their creation in the form of the taking of fifty-two American hostages at the American embassy in Tehran.

As the American hostage crisis and subsequent events proved, the foundations of Khomeini's foreign policy were radicalism, terrorism, blackmail, and hostage-taking. As a result, Iran can count as its friends only a few radical countries such as Libya and Syria. In spite of this, many countries, including most of western Europe and the United States, found ways to do business–especially arms business– with Iran. And, although the Arab countries resented Iran, the fanatical religious leaders in these countries saw in Khomeini a revival of their cause. The most famous victim of this fanatical religious revival was Egypt's President Anwar Sadat.

In shaping a regime under which human life is without value, Khomeini had learned a lesson form the Shah. When the Shah had eased his control over the people, and had given them the political freedom that allowed the Khomeini revolution to take root and grow, he had lost his throne. Therefore, repression, imprisonment, physical punishment and executions were the main pillars upon which the Khomeini regime rested...and survived.

Under Khomeini, every positive aspect of the Shah's regime, such as economic growth, international respect for Iran, personal and political freedom, easy travel to and from foreign countries, education, took an immediate turn for the worse. Yet all the complaints that the radicals had ever had about the Shah's regime intensified dramatically, and reached their apex under Khomeini's stewardship.

The Shah's SAVAK was replaced by Khomeini's SAVAMA. Even the Shah's portrait depicting him with his hand raised in front of his head was replaced by a picture of Khomeini in a similar pose. The transformation was complete, yet neither the Iranian people nor the world gained from Khomeini's replacement of the Shah.

With Khomeini dead, can the mullahs continue to dominate Iran? With the mullahs maintaining firm economic and political control in present-day Iran, the questions in Iran center on which particular mullah and his followers will prevail. The post-Khomeini power struggle will continue, but it will probably make little difference to the Iranian people which faction of mullahs ends up on top. Still, the days of the mullahs are numbered. If they loosen their tight control the slightest bit, the people's revenge will sweep them away. If they continue their oppression of the people, the people will crack under the pressure of such outrage, and the mullahs must fall eventually.

Shortly after Khomeini came into power, the Iranian people began to have their regrets. On walls throughout the country appeared these words: *Marg bar mar, ke goftim marg bar Shah!* ("Death to us, that we said 'Death to the Shah!'" Right then, the stage was set for the downfall of the mullahs and their brutal regime. The real revolution is yet to come.

5

THE STRUGGLE TO CONSOLIDATE POWER

S addam Hussein's strategic, and probably fatal, blunder in his invasion of Kuwait on August 2, 1990, was that he didn't sweep across Saudi Arabia and occupy that country as well. Had he done so, and secured his position by placing explosive charges at critical points in the main oil wells and related facilities of his conquered lands–to be detonated in the event of foreign interference–he would now be in control of a decisive portion of the world's oil supply, and could maintain this position indefinitely. Fortunately, he stopped short of the ultimate victory and left himself vulnerable to the forces of the West.

Although Khomeini lacked the vision to realize the potential of this scenario when the Arabs and Iraqis offered peace in 1982, and Saddam Hussein was short on determination, no one can be sure that a third opportunity would not be disastrous to the interests of the West. If there is a next time, it will likely be the Iranians and fundamentalist Islam provoking the peace.

How this could happen is not difficult to envision, particularly since the invasion of Kuwait and the buildup of U.S.

and other military force in the region. Here are four possible scenarios:

1. The combination of economic embargo and military pressure could force the Iraqis to withdraw from Kuwait. Things could then return to *status quo ante*. Although this is the dream solution for the West, chances of it coming to pass are very slim. Should it occur, Saddam Hussein's days would be numbered, if he could survive the retraction in the first place. He can no more accept this total defeat than the Americans can accept political dialogue without the Iraqis withdrawing from Kuwait.

This scenario would not be as ideal as most Westerners would imagine, anyway, because it does not resolve the fact of Iraqi military might. With nearly six percent of his population at arms, Hussein could begin a replay at any time. Of course, the West could defuse the Iraqi military in the same way the Shah's might was thwarted, by installing a government of mullahs. This would be even worse than leaving Hussein in command.

2. The Americans may finally decide to attack the Iraqis and wipe out their military forces and resources, despite the hostages, whether in an all-out war or by coordinated pre-emptive strikes. This would be a costly move at best, but would likely succeed given sufficient resolve on the part of the U.S. However, this would leave the delicate problem of installing a new government in Iraq. Under these circumstances, anti-American and anti-Western sentiments would dominate the population, boding ill for the stability and durability of a puppet government. While the elimination of Iraq's military would solve some economic problems for that country, the result would be that sooner or later, even if through democratic elections, Iraq would fall into the grips of Islamic fundamentalists.

3. It is possible that war will not break out, and that the embargo could also be ineffective, especially if the Iranians come to Iraq's aid. Remember that, upon invading Kuwait, Saddam Hussein suddenly acceded to all Iranian demands for the return of territory occupied by Iraq since the Iran-Iraq war and for the repatriation of prisoners. This, of course, freed up a large number of Iraqi troops for other duty and took the heat off Hussein's Iranian flank. But no one should discount the possibility of the Islamic Republic providing aid to its former foe. After all, this is the Middle East, where ever-changing truth is stranger than any fiction.

This situation could actually result in the defeat of America and the West, and while it might not necessarily bring about belated military action, it would at least require a prolonged military presence in the region on the part of the U.S. and its allies to guard the resources of Saudi Arabia, essential to the economic survival of the West. This would be very costly, both in economic and political terms, to the West. For the Iraqis it would mean the weakening of their military power and the erosion of their economy to the point of destruction. The winner in this scenario would be the Islamic Republic of Iran.

4. The Americans and their allies could be humiliated in a military showdown with Hussein, and be forced to back out of the Middle East. The implications in terms of economic prosperity and political stability in the West don't need to be spelled out. How likely is this scenario? How likely is it that Hossein's firepower could stand up to an onslaught from the West? Plainly the United States cannot stand for such an outcome. But what if attacking Hussein means certain death for the thousands of Western hostages shielding the military and industrial facilities in Iraq?

Regardless of what unfolds in the Middle East in the near

future, the era of cheap oil is past. Already prices are escalating. Even if the Saudis and others are able to ameliorate the price level by boosting production, the resulting plateau will be much higher than before the invasion of Kuwait. Meanwhile, for the foreseeable future, the overhead of an expensive military presence in the region must be considered in the true cost of oil, and this will likely be so for years to come. And all of this plays directly into the hands of the Islamic fundamentalists, under the leadership of the Islamic Republic of Iran. There can be little doubt that this time around the Islamic Republic will have a deeper appreciation of the potential power of oil.

To understand the situation better, one must consider the background.

The goal of the Islamic Republic is to spread its Islamic revolution throughout the Muslim countries of the Middle East. The first target of the Islamic Republic was Iraq. Khomeini and his fellow mullahs called for the Iraqis to depose Saddam Hussein and his government and form an Islamic Republic. Hussein responded with military force, and the war began.

The war between Iran and Iraq overshadowed the Islamic Republic's other efforts to spread its Islamic Revolution, notably against the Saudi monarchy, and possibly postponed some other plans as well. Despite Iran's lack of success in the war with Iraq, the West should not discount the threat posed by the Islamic Republic and fundamental Islam. While the West stands idly by, the stage could be set for the Islamic Republic and the fundamentalists to grab control of the Middle East by the middle of this decade. Although this outcome is not inevitable, the potential certainly exists, and the implications for the Middle East, and the world, are sobering.

In the latter half of 1988, a tenuous cease-fire was reached in the Iran-Iraq war. The fact that the war had gone so badly had less to do with the prowess of the Iraqis than the ineptitude of the mullahs in the Islamic Republic. The mullahs continued to alienate the rest of the world, while Iraq had the support of the Soviet Union and much of the West. The mullahs did not rely on the strategy and expertise of well-trained military men, but waged war under the direction of mullahs and the Islamic Revolutionary Guard. Later in the war, the mullahs newfound pragmatism came too late.

Although a massive movement such as the full-scale war with Iraq has been unsuccessful, there are many possible events outside of Iran which would work greatly to the Islamic Republic's advantage and further its goals. Among these would be the removal or assassination of Iraqi President Saddam Hussein by military coup, or through the actions of Shi'ite extremists. This would be quite in concert with Iraq's recent history, and not at all unusual in the Middle East. Or a covert action on the part of the West could accomplish the same end, and attempt to credit Iraqi moderates, or extremists, for that matter. Another boost for the Islamic Republic would be the replacement of Egypt's moderate government by either a Ghaddafi-type military dictatorship or simply religious extremists. If Egypt fails to improve its troubled economy or the West stops pumping financial aid into Egypt, the moderate government's days are certainly numbered.

The possibility exists that the cumulative effects of the Islamic Republic's almost routine subversive actions against the Saudi monarchy may bear fruit and advance the revolutionary cause. Iran's intentions toward the Saudis have been well known; witness the bloodshed in Mecca in July, 1987. The United States military presence in Saudi Arabia does nothing to enhance the image of the Saudi monarchy in the eyes of extremist Muslims.

If any of these events occurred, the Islamic Republic's ideological influence would spread deeper into the Arabian Peninsula, menacing Saudi Arabia, Iraq, Kuwait, Bahrain, the United Arab Emirates, Oman, Qatar and Yemen. These countries could become satellites of Iran.

Iran's primary asset in the spread of its influence is the overwhelming majority of Shi'ites in the region. Statistics compiled in 1984 show that of the sixty-odd million citizens of the Gulf states, nearly 46 million–or 75 percent–were Shi'ites. The Shi'ites are distributed as shown in Table 1.

Table 1

SHI'ITE POPULATION IN THE PERSIAN GULF REGION, 1984 [67] (in thousands)			
Country	Total Population	Citizen Popuation	Number of Shi'ites	precentage of Shitie citizens
Qatar	225	70	11	16
Bahrain	360	240	168	70
Oman	950	700	28	4
UAE	1,100	250	45	18
Kuwait	1,370	570	137	24
Saudi Arabia	8,500	5,500	440	8
Iraq	14,400	13,500	8,100	60
Iran	42,000	40,000	36,800	92
Totals:	68,935	60,830	45,792	75%
(Note: According to Iran's 1987 figures, the population of Iran has exceed fifty million.)			

Iran's position is further enhanced by the presence of over 300,000 Iranian Shi'ite emigrants in the Gulf states of Kuwait, Bahrain, the United Arab Emirates, and Qatar. These people could represent a significant extension of the Islamic

Republic's ability to foment unrest or revolution beyond its borders.

In spite of the vast amount of financial aid bestowed upon Iraq by the Arab nations, and the shipment to Iraq of huge supplies of arms from a variety of sources (notably the Soviet Union), and despite the embargo on arms sales to Iran, the Islamic Republic has survived in the face of such enormous odds, and can probably continue to do so. The Shi'ite population in Iran outweighs the total population of the entire rest of the Gulf region.

Should the fundamentalists, under the leadership of the Islamic Republic of Iran, manage to control the oil-rich Gulf region, it is highly unlikely that the superpowers of the West would band together militarily against the whole region as they did against Iraq. It is equally unlikely that the United States and Western Europe would use military force to prevent the Islamic Republic from gaining control over the Arabian Peninsula in the first place.

As we have seen in the recent confrontation with Iraq, the preponderance of the military burden has fallen to (or been taken on by) the United States. The reason that the Western European countries would not get involved in a conflict of this nature is not simply that they are so dependent on Middle Eastern oil. According to Helmut Schmidt, the former chancellor of West Germany, the Western European countries have neither the military might nor economic power to sustain a war effort in this region.[68]

The most that the United States has been able to achieve in the region so far has been to use its navy to show off, escort oil tankers and blow up a few unoccupied oil rigs in the Gulf. The Americans, like their European allies, are ill-prepared to sustain a conventional war. Further, the Americans would have to overcome the psychological problems they have suffered as a nation since Vietnam, which have been exacerbated by the deaths of over 240 U.S. Marines at the hands of Iranian suicide commandos in Lebanon in 1983, after which the Americans went home, and by the accidental downing of an Iranian civilian airliner by U.S. naval forces in the Persian Gulf in 1988. These images, superimposed on those of Americans scrambling on the roof of the U.S. Embassy in Saigon, hanging from helicopters to evacuate in the face of the advancing victorious North Vietnamese, do nothing to strengthen national resolve. Add to that thousands of strategically-placed Western hostages in the region, and friendly nations threatened by chemical weapons, and there is indeed a lot of resolve needed before decisive action can be taken.

Another sobering deterrent to U.S. military involvement in the Middle East is the devastating example of the Soviet occupation of Afghanistan, and the lessons to be learned from their experiences there. The failure of the mighty Soviet Union to subdue the outnumbered and poorly-equipped Afghanis, despite the geostrategic and military advantages of the Soviets, does not bode well for any similar activity on the part of Americans in the Middle East.

In the new climate of U.S.-Soviet relations, it is unlikely that the Soviets would provide military aid to the Islamic Republic. However, there could be the threat of involvement on the part of the rest of the Muslim world, including Pakistan, Afghanistan and Libya. These countries could well mobilize to show solidarity with the Islamic fundamentalists against the Great Satan (as the U.S. is known in Iran). Certainly the longer the U.S. military presence in the region, the more anti-American sentiment will accrue. And if the United States were to attack and occupy Iraq, for reasons however noble or pragmatic, resentment would inflame the fundamentalists more than ever.

While Soviet support of the U.N. Security Council resolution sanctioning military force to enforce the economic blockade against Iraq is welcome it is unlikely that the Soviets will play an active role in the situation. The Soviets have gained, and stand only to gain more, by taking a passive position. However the situation resolves itself, the price of oil will remain high, much to the benefit of the cash-hungry Soviet economy. Besides vodka, diamonds and gold, the Soviets export oil.

It is doubtful whether any outside power could intervene successfully against an Islamic revolution spreading into Iraq or across the Arabian Peninsula. The Islamic radicals always have the terrorist trump card: death commando suicide raids on military installations, or simply on the oil wells, refineries and pipelines which could cut off the vital supplies of oil and bring the West to submission. A few hours' work could cut off the supply of oil for years.

Logistically, the oil wells, pipelines and refineries are extremely vulnerable. Further, there is a high density of Shi'ite population around the major oil fields. The eastern province of Saudi Arabia has a Shi'ite population of approximately 450,000, and the world's largest oil field is situated in the al-Hasa district, nearly 60 percent of whose population are Shi'ites.[69]

In addition to the Shi'ites, there are other Muslim extremists throughout the Arab countries who desire either to impose full Islamic law, or, better yet, to replace their secular governments with Islamic theocracies. Since the Islamic Republic of Iran is an inspiring example for Muslim radicals, the full added support of these groups could be expected by Iran. An example of the terrorist exploits of one of these groups was the assassination of President Anwar Sadat of Egypt during a military parade in October, 1981, by the radical Muslim underground group Jihad. The same could well happen to any leader in the Middle East–or elsewhere.

Americans and Western Europeans tend to imagine that if the Islamic Republic of Iran were confronted with the military might of the West, it would be intimidated and refrain from its radical course of action. This only demonstrates the West's lack of understanding the facts. The threat of military force does not intimidate radical Muslims, especially not radical Shi'ite Muslims. To understand why this is true, one only needs to look to the *Koran*.

According to the dictates of Islam, as set forth in the *Koran*, anyone who is not a Muslim is classified as *kafir* (infidel). It is the duty of the *moemaen* (Muslim believer) to convert the infidel to Islam by whatever means necessary, however radical or violent. To the true believer in Islam, the attainment of heaven is the ultimate goal. According to the *Koran*, if a believer kills in the name of Islam, or is killed in combat in a holy war, he is promised heaven. Because of this promise, and because death in the course of battle achieves martyrdom, the most glorious status a Muslim can achieve, and because of the illiteracy of the vast majority of Muslims, the threat of military force does not frighten the radical Muslims; it more likely excites them.

While the West indisputably possesses far more sophisticated weaponry than does the Islamic Republic, the West is very poorly equipped for a war of nerves, cruelty and self-sacrifice. If the Americans were demoralized by Vietnam, one must wonder what the reaction of the Western media, people–and the military themselves–would be if the West were confronted by an army of children, armed with rifles and grenades, and ready to die for Islam, as was sent against Iraq. Would the West be able to shoot, and keep on shooting until the last child was martyred?

Should the Islamic Republic consolidate its control over Saudi Arabia, Kuwait, Iraq, Bahrain, the United Arab Emirates, Oman and Qatar, Iran would need access to the Mediterranean Sea for geostrategic reasons. This could be obtained by annexation of Syria and Lebanon. Bringing these two countries into the fold would not be difficult. Iran already has a strong foothold in Lebanon in the *Hezbollah* (Party of God), a terrorist organization founded, directed and financed by the Islamic Republic of Iran. In Syria, the potential power is in the hands of the Alawites, a secretive sect of Islam leaning towards the Shi'ites. At present, both Syria and Lebanon are pro-Iranian.

Although Jordan is a country without strategic or economic significance for Iran, it may well be possible that Jordan would prefer to come under the umbrella of the new superpower regime and benefit economically, rather than to be an outcast in a region dominated by the Islamic Republic. In recent years, there have been signs of the growing influence of fundamentalist Islam in Jordan; in 1989, the fundamentalists captured over 30 seats in the Jordanian parliament.

Should Iran achieve its goal of expanding the Islamic Revolution to envelope the Gulf region, the new boundaries of the Islamic Republic's sphere of influence would be the Soviet Union in the northeast, as before, with new lines in the west at Israel and the Mediterranean Sea, in the southwest at the Red Sea, and in the south at the Arabian Sea. The map following shows the Islamic Republic's potential area of domination.

Should the Islamic Republic not achieve a position of dominance in the Middle East through force, the very necessity to rebuild the economies of Iran and Iraq could lead to a consolidation of power in the entire region in the hands of the Islamic Republic, through peaceful but no less effective

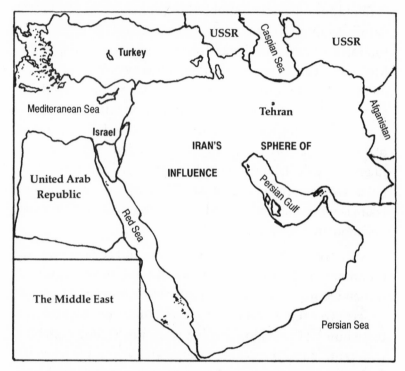

This map illustrates the potential sphere of influence of the Islamic Republic should the Islamic Revolution continue to spread.

means. Thus could come to pass the third scenario outlined at the beginning of this chapter.

Iran and Iraq both desperately need to rebuild their economies, which have been practically destroyed by the war. Their first step will be to pump more oil to get more income. But the world oil market, as it stands, will not absorb more oil without further depressing prices. The situation could get even worse if other oil-exporting countries increased their output levels to raise income, which would further weaken prices. The short-term benefits of increased oil production would be marginal at best, with long-term consequences that would more than offset any immediate gains.

The remedy for Iran, Iraq, and all other oil-producing nations, can only be found in a coordinated oil policy, as will be described in Chapter Eight.

In order for the Islamic Republic of Iran to dominate the Middle East through economic rather than military means, the Islamic Republic would have to take the lead in the Organization of Petroleum Exporting Countries (OPEC). While Iraq's interests in greater oil revenue coincide with the Iran's, Iran is in a much better position to take the initiative in establishing new policies for its own benefit, which would benefit all the oil-producing countries. This is especially true now that Iraq is backed into a corner. Iran has far greater oil resources than Iraq, when Iran's "off-the-books" oil resources are considered ("known reserves" often reflect the oil companies' policies rather than reality).

Iran's new oil policy would be challenged at first, particularly by the Saudis, who in the past have been more considerate of the interests of the West, notably those of the United States, and are now more so than ever. Before long, however, the Saudis and all the oil-producing nations of the Middle East would come to support, at least tacitly, the Islamic Republic's oil policy, for three reasons. First, they would not want a conflict with revolutionary Iran; indeed, as time went by, they would be more desirous of the Islamic Republic's good will. Second, they would prefer to see Iran and Iraq rebuild their war-ravaged economies rather than to rekindle and continue the war. And, most importantly, if the direction of Iran's oil policy followed the scenario outlined in Chapter Eight, it would be extremely beneficial to all oil-producing nations, the Saudis included.

The consolidation of Iran's control of the Persian Gulf area would catapult the Islamic Republic into a position of economic and political power with worldwide ramifications.

With Iran controlling or influencing 84 percent of the world's oil surpluses, a drastic shift of economic and political power would immediately occur in Iran's favor. And then, for the first time in history, an oil-exporting country would be the leading monetary power in the world, setting the course of the world's monetary system. The Iranian rial would replace the U.S. dollar as the world's most desirable currency; the German mark and the Japanese yen would also lose their luster.

Today, one may easily underestimate the economic and political potential of the Islamic Republic, and their worldwide implications. However, one may also recall the position of Iran under the Shah prior to 1970. Iran was known in the West as a backward, underdeveloped country, whose most significant achievement was its beautiful rugs. The Shah himself was best known for his wives, especially Queen Soraya, his second wife. Until the beginning of the 70s, Iran and the Shah were presented by the boulevard-press in the West as fairy tales.

After 1971, when the price of oil began to increase dramatically, the Shah emerged as a statesman and the leader of an economically and politically significant nation of the world. By 1975, the Shah and Iran had reached the height of their glory.

The difference between Iran's emergence from obscurity in the 1970s and what could occur in the 1990s is that the rapid rise of the Shah and Iran in the 70s took place under circumstances far less favorable to Iran than those of the 90s could be. In the 70s, Iran was an important but minor member of OPEC, with comparatively little to say about oil production and pricing policies.

In the 1990s, the Islamic Republic of Iran could rise much farther and much faster to the status of an economic and political superpower.

6

THE DESTRUCTIVE POWER OF THE ISLAMIC SHI'ITE

*T*he most decisive factor which could tip the scale in favor of Iran's ultimate dominance of the Middle East is the destructive power of the Islamic Shi'ite. Islam is the most radical religion known to man. Of the two largest sects of Islam, the Sunnis and the Shi'ites, the Shi'ites are by far more extreme. The Shi'ites make up over 75 percent of the population of the Persian Gulf region, and Iran is their main homeland, where over 90 percent of the population is Shi'ite.

Although there is growing awareness of Islam in general, and the Shi'ites in particular, very few people comprehend the extreme views of this religion. One can hardly blame the non-Muslim world for its lack of knowledge of the tenets of Islam, but it is astonishing to discover that the adherents of Islam, including its ecclesiastics and intellectuals, are ignorant of its teaching, and clamor in a sea of know-nothing-ism. However, what the Muslims lack in knowledge of their religion, they more than make up for in fervor; their zeal often becomes violent. This is a common characteristic of Muslims, Sunni, Shi'ite, or otherwise.

The violent fervor of the Muslims, in and of itself, indicates that its followers have been thoroughly brainwashed by the constant preaching of the mullahs. The Islamic Republic's violent character and terrorist adventures are, in fact, manifestations of Shi'ite Islam as it is practiced by the mullahs: exactly the Islamic Republic the mullahs say it is. Iran has made the nature of Islam more visible, if no less comprehensible, to the non-Muslim world.

In the years since Khomeini's revolution in Iran, more books have been published in the Persian language criticizing Islam and the *Koran* than in all previous times combined.[70] In studying these books, which are well-versed in the tenets of Islam, the impression is that the Islamic religion has been like a pot full of rotten brew, and that Khomeini's accomplishment was to lift the lid from this pot...letting the smell spread around.

An understanding of Islam is essential to understand the thesis of this book, of Iran, the Middle East, and what is going on in the world today. The presentation of Islam here is far from a complete and exhaustive study of this interesting phenomenon. The intent here is to present a well-rounded overview, based on original sources and the works of Islamic scholars.[71]

The Prophet Mohammed,[72] the founder of Islam, was born in 570 A.D., a few months after the death of his father, Abdullah. When Mohammed was only six years old, his mother, who had been mentally and physically ill, died. Mohammed lived in total poverty during his youth, and suffered immensely from it. This is the main reason for his insistence, in no less than twenty-three parts of the *Koran*, that orphans should be helped and their rights defended.

The years after his mother's death were even tougher on Mohammed. He went to live with his grandfather, Abdul-

Motleb, who was 108 years old at the time. Mohammed's grandfather died in 579. After his grandfather's death, Mohammed's uncle, Abu-taleb, a merchant, became his guardian. Under his uncle's tutelage, Mohammed became a camel driver. Between the ages of ten and fourteen, Mohammed traveled several times between Mecca and Syria. The final leg of his return from his last journey coincided with the time that his uncle was gathering men from the various tribes in order to defend themselves against an imminent attack by the ruler of Abyssinia (Ethiopia).

Young Mohammed was expected to participate in the war, but since he was very weak and nervous he escaped this duty. Mohammed's behavior caused ridicule from his friends and acquaintances. As a result, he left his uncle's house, at the age of twenty-five, in 595. To earn his living, he was forced to become a shepherd. Mohammed suffered greatly in his new calling, because shepherding was the lowliest of trades. Soon, Mohammed went to work for a cloth merchant named Saheb, whose place of business was in one of the famous bazaars of Mecca. It was here that Mohammed made the acquaintance of the rich widow Khadijeh.

Khadijeh employed Mohammed, who became her partner and eventually her husband. Although Khadijeh was fifteen years older than Mohammed, the age difference did not seem to bother either of them. Mohammed was happy and grateful to Khadijeh for marrying him, because this put an end to his misery. Khadijeh, who was forty years old, was delighted to have a much younger man for her husband.

Until his marriage to Khadijeh, Mohammed was a relatively unknown person who had no prior intention of becoming any sort of prophet or founding a new religion. Since nothing is known of Mohammed's education, the Muslims believed that he was illiterate and thus considered his revelation of the

Koran miraculous. Therefore, it was postulated, the *Koran* was the word of God revealed to Mohammed, and as such was infallible.

Prior to Mohammed's declarations and the initiation of Islam, there were groups of Jews who lived in Hejaz (Arabia). Since these Jews were in contact with Jews from all over the Middle East, they frequently had visitors from abroad. It has been postulated that, in order to maintain secrecy in all matters, Mohammed received his education from foreign Jews. This would seem to explain why there are so many verses in the *Koran* which show a strong influence of the *Old Testament*.

Until the age of 35, Mohammed concentrated on expanding his wife's business and on preparation for prophethood, for which the silence of the desert and the mountains provided the perfect backdrop. Due to his success in business, Mohammed amassed considerable wealth. By his thirty-fifth birthday he was considered a wealthy man in his own right, and led a comfortable life.

However, Mohammed still remembered the painful times of his childhood, and the poverty and hard times of his youth. As a result, wealth did not beget arrogance. On the contrary, he was more sympathetic to the plight of the poor, and avoided the wealthy people of Mecca. The people of Mecca had little esteem for Mohammed, either, since they considered his wealth to be due to his marriage and not from his abilities. Becoming rich by marriage was traditionally taboo.

Mohammed's marriage to Khadijeh gave him peace of mind, and the time and comfort to develop his thoughts into a brand new religion. During this time, his thoughts centered on subjects with which he was more or less familiar. He recalled, for example, the poverty of his youth, his suffering as a shepherd, the cruelty of the wealthy people of Mecca, and the differences among the tribes; all of these and more guided his train of thought.

In addition, the preachings of the Jews of Medina that a messenger would come who would bring peace and justice set quite well with Mohammed's thinking. If he were to start a new religion, not only could he lead the people of Mecca to forsake idolatry, but he could unify all of the people of the world under one banner. To facilitate the conversion of Jews and Christians to Islam, Mohammed introduced himself as a follower of Abraham, who had been neither Jew, Christian, nor idolator, but a true believer who had worshipped God. Mohammed thought that since the Jews and Christians accepted Abraham as a prophet of God, they would be eager to accept Mohammed and join the ranks of his followers. So, at the commencement of his prophethood, Mohammed did not oppose any of the previous religions; he simply invited their followers to accept Islam, which he presented as a continuation of Judaism and Christianity.

It did not take long for Mohammed to realize that he could never expect any support from Jews or Christians, so he decided to introduce Islam as an entirely new religion. As a result of this decision, changes were made in the *Koran* to accommodate changing circumstances. Finally the foundation was laid for the emergence of a new religion which could compete with Judaism and Christianity to such an extent that today Islam boasts over 850 million followers worldwide.

Mohammed's prophethood began rather simply. One day he returned from the cave of Hara and told his wife that an angel named Gabriel had appeared and told him, "You are the prophet of God, and I am Gabriel." After three years had passed, Gabriel again appeared to Mohammed and told him, "You are the prophet of God, and I am Gabriel." This three year period between Gabriel's appearances is designated the

Doure fetrat (period of interval) which Mohammed needed to prepare himself spiritually for the great undertaking of prophethood. This explanation seems logical, because the program he designed during that time helped him tremendously for the next twenty-three years and, as the world has seen, his prophethood has since been proclaimed worldwide. When Khadijeh heard of Mohammed's meeting with angel Gabriel and that Mohammed was pronounced God's prophet on earth, she told her husband that she was sure that he was a prophet and that she believed in him. Mohammed had his first follower.

The second person to declare himself a follower of Mohammed was Zeid, Mohammed's slave. Mohammed's third follower was his sixteen-year-old cousin Ali, the son of Abu-taleb. Ali later became Mohammed's son-in-law. According to Sunni tradition, Ali was the fourth caliph of Islam (the caliph being the leader of Islam after Mohammed's death); according to the Shi'ites, Ali was the first caliph. At age sixteen, Ali was ready for excitement and adventure; he was to find both in the years ahead.

Despite his three followers, Mohammed's attempts to attract more to his fold proved fruitless, and during the next three years he managed to increase his flock to only thirteen. To break the impasse, Mohammed decided to invite the forty heads of the Qoreyshite tribe to which he belonged to a meeting, where he would declare himself the prophet of God and founder of Islam, and invite them to embrace this God-sent religion and to pledge their allegiance to him. Not only did the forty chiefs reject his invitation, they were hostile toward Mohammed, primarily because of his disrespect for their idol temple in Kaabe.

Mohammed then turned to the strangers who traveled to Mecca, especially those from the city of Medina, to attract more believers. The travelers from Medina supported his claims, primarily for two reasons: first, they were receptive to such claims, because the Jews had been promising the appearance of a new prophet; and second, the teachings of Mohammed proclaimed the existence of One God, which was an attack upon the idols of the people of Mecca. The people of Medina, who were long-time rivals of the people of Mecca, found in Mohammed a chance to defeat their mutual enemy, the people of Mecca. The more hated Mohammed was in Mecca, the greater his stature in the eyes of the people of Medina.

The people of Mecca's hatred of Mohammed escalated to the point that they wanted to expel him forever or, better still, to kill him. The mediation of his uncle, Abu-taleb, who believed that Mohammed was mentally ill, prevented the Meccans from taking such drastic steps.

In the year 619, Mohammed's uncle died, and a year later Khadijeh also died. At this time, Mohammed began in earnest to seek followers. Since he knew that he would get no support from the people of Mecca, he began to negotiate with the people of Medina. The people of Medina welcomed Mohammed into their city, and Mohammed declared himself to be of the same blood as the people of Medina.

When the people of Mecca heard of this treachery, it was decided to slay Mohammed. Mohammed escaped to Medina in September, 622, with a few of his followers: Zeid, Ali, Abu-bakr (his new father-in-law and the first caliph according to the Sunnis), Omar and Osman (the second and third caliphs of the Sunnis). This flight is known in Islam as the *Hejrat* (migration).

If the people of Medina had not embraced Mohammed, perhaps Islam would have died with him. As it was, the escape to Medina marked the start of the rapid expansion of

Islam. It could be said that Islam was born out of the rivalry of the peoples of Medina and Mecca. To be precise, the people of Medina accepted Islam more as an anti-Mecca movement than as a religion of divine origin.

Upon his arrival in Medina, Mohammed built a mosque as a counter-balance to the Meccans' idolatrous temple, the Kaabe. His next move, although he was ill-prepared for it, was to declare war on Mecca. The war was declared for two reasons: the desire of the people of Medina to get even with Mecca, and Mohammed's desire to crush the Meccans' unceasing ridicule of him. But there were further incentives for the war. If Mohammed could defeat Mecca and convert the people to Islam, his soldiers would be entitled to the loot and spoils of war; and, if he were victorious, he could claim that his triumph was a sign from God and a miracle—proof to his followers of his legitimacy, which was long overdue.

During this period, the foundations of Islam's radicalism were laid, for which the world has suffered ever since.

Mohammed began his attacks on Mecca in 624. In the first skirmish, the Battle of Badr, Mohammed was victorious and gained followers. In the next confrontation, the Battle of Ahad, however, he was defeated by the tribe of Qoreysh. After that, Mohammed took on the smaller and less powerful tribes, gaining easy victories, more followers, and plenty of spoils for all. Meanwhile, Mohammed's soldiers were becoming experienced fighters. Finally, on January 12, 630, Mohammed conquered the city of Mecca.

During the plundering of Mecca, Mohammed's troops destroyed sixty idols in the Kaabe, which represented the gods of the sixty tribes. He declared the Kaabe the house of the One True God, Allah. The defeat of Mecca added to the glory of Islam, and by 632 the whole of Arabia embraced the new religion. In the same year, Mohammed entered Mecca with 40,000 of his followers to celebrate the victories and expansion of Islam. In a speech he asked, "O Great God, did I complete the prophetic mission, which I was chosen to do?" The audience answered, "Yes, you completed your prophetic mission!"

On the return to Medina from this celebration, Mohammed fell ill and died in the arms of his most beloved wife, Ayeshe, on June 8, 632.[73]

Mohammed's career differed from that of the classical religious prophet; he was more like the leader of a political party. His main goals were personal power, through converting people to Islam, and material wealth, through the spoils of holy war; both are endorsed by the *Koran*. Islam, which Mohammed promoted with the brutal power of the sword, is certainly unique among religions.

Islam does not tolerate non-Muslims or adherents of other faiths. All non-Muslims are considered "infidels." The followers of Islam have divided the world into two parts: Dar-ul-Islam (House of Islam) which includes those parts of the world where Muslims live, and Dar-ul-Harb (Place of War), those parts which are not yet conquered by the Muslims.

According to the Muslims, the two parts of the world can never coexist in peace, until the Dar-ul-Harb comes under the wings of Islam. The leaders of Islam may live in peace with the non-Muslim parts of the world, but it remains the duty of each Muslim to conquer the non-Muslim parts of the world and bring them under the control of Islam. Therefore, it is the obligation of each Muslim to take part in the *Jihad* (holy war) to conquer the infidels.

Any Muslim who kills a non-Muslim, or is killed by one, in the Jihad will be considered a martyr and will go directly to heaven and be blessed by God, according to verses 154,

169 and 170 of the Sura (chapter) Baghara (ox) of the *Koran*.

By implementing the concept of Jihad, Mohammed succeeded in eliminating his enemies. The same attention to principles enabled his followers to spread Islam from Arabia to the borders of China in the east and Spain in the west by 750. Mohammed left no room for misjudgments in the proper manner of dealing with infidels. In verse 4 of the Sura Mohammed, Mohammed commands the Muslims: "Therefore, when you come face to face with the infidels, chop off their necks, so that the flow of your enemies' blood brings them to their knees in submission, thence bind the captives with ropes, so that you can demand ransom for their release."

In Islam, taking bounties and spoils from the vanquished, and raping conquered women, are allowed. Mohammed included these features in his design for Islam because in the early days he needed money to finance his holy wars, and these incentives were also very useful recruitment tools for his army. Soldiers are easily motivated by loot and women, and Islam offered both. Verse 24 of the Sura Nesa (woman) in the *Koran* says, "Sexual involvement for you with married women is prohibited, unless you have taken these women from the vanquished infidels as spoils of war."

Although Mohammed allows his followers the spoils of war and ransom for captives, he requires that one-fifth of these gains be given to God, the Prophet, and his descendants, as well as to the poor and orphans, as commanded in verse 41 of the Sura of "Enfall".

The twenty per cent duty of Islam has been a harsh claim on the wages of hard-working people ever since, simply because once upon a time Mohammed made the taking of spoils and ransom legitimate, in order to expand Islam.

In Islam, the social status of women is low to nonexistent. For example, in verse 34 of the Sura of Nesa (woman)

Mohammed says, "If your wife disobey
you, in the first instance counsel her. If th
her of her conjugal privilege. If she stil
your wishes, you may beat her up." Si
allowed four wives, it is doubtful that he w~ ~~
he has to straighten one wife out by denying her sexual favor.

According to some interpretations of Islam, beating the
wife should stop short of breaking her bones, since that could
be considered a criminal act.[74]

It is interesting to note that God took a personal interest in
the sex life and marriages of Mohammed, who tells us God's
words in verse 50 - 53 of the Sura of Ahzab (parties):

"You, O prophet, in sleeping with your wives, it is not
incumbent upon you to observe a sequence. Whichever of
your wives you wish, you could postpone their turns, and
whichever of them you put aside, you could have them back
later. It should make them happy when you take them back
and none of them should ever become unhappy or displeased because of your desires. On the contrary, all of them
should be happy for what you desire."

It is a good thing that God offered this advice, because
Mohammed had twenty-five wives. All his wives were
widowed before he married them, except for Ayeshe, who
was only six years old when she became Mohammed's wife.
Ayeshe was the daughter of Abu-bakr. She was forty-five
years younger than Mohammed; when he died at 63, she was
18.

According to the *Koran*, a woman is not to be judged by
her education or achievements, but by her obedience and
servitude to her husband. By excluding women from the
world of business, the Islamic world deprives itself of the
contributions and productivity of half its population. A

primary reason for the economic backwardness of the Islamic world is its traditions concerning women.

In fairness it must be mentioned that the entire *Koran* is not as radical as the above excerpts. Actually, there are two main parts of the *Koran*. The first consists of the suras which came from Mohammed in the early days of Islam in Mecca. These suras present Mohammed as a prophet who preached the virtues of humanity, peace and right living to his followers. This part of the *Koran* was written when Mohammed had no power, when he and his followers had no acceptance in Mecca, which is the main reason for the modesty of the rhetoric in this part of the *Koran*.

The second part of the *Koran* was written during the time when Mohammed was living in Medina, seeking to conquer Mecca and the rest of Arabia. These suras depict Mohammed as a religious crusader and conqueror who is merciless and vengeful. Radicalism became the trademark of Islam at this time, and continues so to this day. However, events which took place after Mohammed's death developed the radicalism of Islam even further. Iranian and non-Iranian Muslims often blame Khomeini and his followers for ruining the reputation of Islam and causing embarrassment worldwide for this peace-loving, gentle religion. Although Khomeini did nothing to enhance Islam's image, the Muslims ignore the fact that half of their *Koran* teaches killing the infidels, and taking hostages for ransom.

Mohammed's death in 632 set into motion forces which would lead to Islam's cleavage into two main branches and countless smaller sects. The two main branches are the Shi'ites, who are the majority in the Middle East, and the Sunnis, who outnumber the Shi'ites many times over globally. The main cause of the split was the dispute over who was going to succeed Mohammed. To this day, over 1,300 years later, a satisfactory answer has not been arrived at. The learned scholars of Islam are at a loss to explain why Mohammed failed to choose his successor when he was alive. After all, Mohammed meticulously attended to every detail of even the most mundane matters, as shown in the *Koran*. Had he appointed a successor, generations of Muslims might have spent more time fighting the infidels than they did each other.

In the days following Mohammed's death, some Muslims thought that Ali Ibn Abu-taleb, Mohammed's cousin and brother-in-law, should succeed Mohammed as the leader of Islam and become the first caliph of Islam. Another group believed that Abu-bakr, father of Ayeshe and father-in-law of Mohammed, was more qualified for the position, because he had almost always been with the Prophet, and thus was much closer to him.[75]

In the end, despite disputes and intrigue, Abu Bakr Ibn Ali-Quhafah became the first caliph of Islam, from 632 to 634. Following him were Omar Ibn al Khattab, the second caliph (634-644), Osman Ibn Affan, the third caliph (644-656) and Ali Ibn Abu-taleb, a cousin and son-in-law of Mohammed, the fourth caliph (656-661). This is the order of succession as it happened, and, the Sunnis believe, rightfully so. The Shi'ites, however, believe that the legitimate first caliph should have been Ali, and that Ali was cheated out of his rightful position.[76]

Whichever faction was right, the schism would have been insignificant because there were very few Shi'ites at the time, and Shi'ism might have faded into history had not the Muslims attacked Iran in 642. By 651, Iran was completely conquered by the Muslims and under the government of the Caliph of Islam. The Iranians embraced Islam, but resented the rule of the Arabs over their Aryan land. So the Iranians rallied to the Shi'ite cause, and revolted against the Sunni Arabs in 749, finally driving the Arabs out of Iran, and wresting control of Baghdad and Damascus as well.

Two further incidents heightened the anger and annoyance of the Shi'ites against the Sunnis. The first developed when Osman, the third caliph, was murdered, and Ali was accused of complicity in his assassination. Even so, Ali became the fourth caliph of Islam. The governor of Syria, a man named Muawiyah, did not recognize Ali's authority, and Muawiyah brought his army against Ali, ostensibly to exact revenge for Osman's murder, but also to further his personal ambitions.

Although Ali's army was winning and had the upper hand in Syria, for some reason Ali halted the conflict and had the dispute settled by two arbitrators. When the verdict ruled in Muawiyah's favor, Ali's reign as caliph ended and Muawiyah became the new caliph of Islam. Some time later, as he was praying in a mosque in Kufeh, Ali was murdered.

The second incident which enrages Shi'ites to this day was the murder of Hussein, who was the son of Ali and the grandson of Mohammed. Hussein died at the hands of Yazid, the son of Muawiyah. After Muawiyah's death, Yazid became the new caliph of Islam. The cruelty with which Hussein and his followers were murdered in Karbala makes him the symbol of martyrdom to the Shi'ites. Every year the Shi'ites mourn his death and re-enact the scene of his martyrdom, in the Islamic month of Moharram.

For all their differences, the Sunnis and Shi'ites have a lot in common. Both sects believe in *Touhid* (divine unity), the Prophet, and the *Koran*. The major philosophical difference between the two sects concerns the principles of *Imamat* (leadership). The Shi'ites believe that the leadership of the Islamic community was the exclusive prerogative of the

Prophet himself. Thus, in their view, Ali, the Prophet's son-in-law, should have succeeded as (presumably) willed by the Prophet himself. The Sunnis, on the other hand, believe that the Prophet Mohammed formed an Islamic society based on the *Koran*, and that this completed his mission. The Sunnis believe that after the passing of the Prophet, it was up to the people to elect a leader according to their own judgment. They further believe that if the Prophet had really wanted to nominate anyone for the Imamat, he would have done so and made his choice clear.[77]

Since the Shi'ites believe that Ali was Mohammed's hand-picked successor, they do not recognize the legitimacy of the first three caliphs. According to their theology, there were twelve heirs or successors of Mohammed, called Imams, in the lineage. The line began with Imam Ali, down through ten of his descendants, and ends with the twelfth Imam, Mohammed al-Mehdi, or Imam Mehdi. He is believed to have disappeared into seclusion during his childhood, in 873, never to be heard from again. The Shi'ites believe that the reappearance of Imam Mehdi, the "Hidden Imam", at some unknown time to come will rescue them from tyranny and oppression, ushering in a new era of justice and bliss in human society.

In contrast, the Sunnis neither accept the legitimacy of the twelve Imams of the Shi'ites, nor do they believe that Imam Mehdi will return.

Besides these major points of contention between the Sunnis and Shi'ites, there are some subtle differences in the way they practice Islam. One such difference concerns the *Salavat*, the special formula for praise and greeting of God. When the Sunnis say the Salavat, they address God and Mohammed; when the Shi'ites say it, they also include the descendants of the Prophet, Ali and his progeny. Another difference between the sects involves the number of wives a man may have. Both sects allow men to marry up to four women and still remain within the tenets of the faith, but the Shi'ites also allow the *sighe* (temporary) marriages of as many wives as a man wishes and can afford.

The Shi'ites believe that legitimate authority is vested in the Imam alone. As such, if the sole legitimate successor of the Prophet is no longer on the earthly plane, this means that any worldly power which claims such authority must be illegitimate, unless it can be demonstrated in a clear and indisputable fashion that the authority is exercised on behalf of the absent Imam.[78] This is where the Grand Ayatollahs (Ayatollah al Ozma) come in.

A Grand Ayatollah is a Mujtahid, one who forms his own independent opinions on religious matters. During the period while the Hidden Imam remains concealed, the Mujtahids act as his agents to guide the faithful in religious matters. A Mujtahid who has reached the highest level of Islamic learning, is known for his piety, and commands a following among the people, earns the highest position in the Shi'ite religious hierarchy; he becomes a *Marja-e-Taqlid* (Source of Imitation). A Mujtahid who is a Marja-e-Taqlid is called an Ayatollah al Ozma (Grand Ayatollah).

Generally there are several Grand Ayatollahs at any given time, who, on behalf of the Hidden Imam, have assumed the burden of interpreting Islam. Since the days of the Safavi dynasty, which promoted the Shi'ite sect in Iran, every Grand Ayatollah has interpreted the problems of Islam in books entitled *Touzieh-ul- Masael Ayatollah - - - (Ayatollah - - -'s Interpretations of Problems).*[79] These books provide Islamic guidance on every aspect of personal life, including marriage, cutting one's nails, eating, having sex with animals, for example, as well as religious, social and political matters. The

Grand Ayatollahs prescribe a solution for every problem known to man. Excerpts from Ayatollah Khomeini's *Interpretation*, at the beginning of Chapter 4, have afforded a sample of the thoughts and ideas of the Grand Ayatollahs.

The leadership of the faithful clearly falls upon the shoulders of the Grand Ayatollahs[80], until such time as the Hidden Imam manifests himself as the savior of all mankind. At which time, he will presumably deliver mankind from the Ayatollahs.

With the Ayatollahs having so much say about everything, it was not outside the position of a Shi'ite Ayatollah to speak of Islamic government, as Khomeini demanded in his writings.[81] What caught everyone completely off guard, however, was that Khomeini seized power, he proceeded to implement exactly what he had written.

There is speculation as to whether the earliest edition of Khomeini's book *Hukumate Islami (Islamic Government)* represented an essay by a British intelligence officer and specialist on Iranian affairs, Ann Lambton.[82] Regardless, what is relevant is the fact that the very existence of this book and its ideas of Islamic government received no attention from politicians or so-called experts, until it was too late.

Had anyone, at least those who followed Khomeini blindly and supported him so wholeheartedly, read his book and realized what he had in mind, Iran and the world would certainly have been saved from his madness. Had they taken the time to read *Hukumate Islami*, they would have foreseen all of the events that subsequently have come to pass in Iran and the Middle East, clearly spelled out in type, with no room for misinterpretation.

Here are a few excerpts from Ayatollah Khomeini's *Hukumate Islami*:

Was it not in the time of the Prophet Mohammed that he not only made the laws, but also executed them, by chopping off the hands of a thief? Punishments with lashes from the whip and stoning to death were common for corresponding crimes. The Caliphs are there to carry out these laws, not to create them.

Introduce Islam to the people so that the young generation does not think that the mullahs in Qom and Nadjaf have no interest in politics, and that religion must be kept separate from politics. The stricture that religion must be kept separate from politics, and that the Islamic clergy should not get involved in politics was instituted by the exploiters of our land and the agents of the colonial powers. It is also what the infidels say. Was religion separate from politics in the time of the Prophet? In those days were the clergy composed of one group of people and the politicians and rulers composed of yet another group?

Khomeini explains the necessity of Islamic government this way:

In order to create the unity of the Muslims, and to rescue the Islamic countries from the influence and the colonization of the foreign countries, the Muslims have to topple the rulers of their countries and establish a just Islamic government.

About the make-up and characteristics of such an Islamic government, he says:

An Islamic government in not a dictatorship, nor is it an absolute monarchy. It is more like a constitutional form of government. Of course, the constitution of an Islamic government is not similar to the ones being practiced at present, which are dependent on the majority of the votes for a law to be passed. The constitution of an Islamic government is based on the laws of the *Koran*, which have been established by the Prophet.

If the Iranian people had taken the trouble to read the following quotation, they would have realized that the Islamic government which Khomeini prescribed is ruled only by the *Faqihs* (Islamic theologians):

> "Should the kings be obedient to Islam, they must follow the Faqihs. They must also inquire of the rules and the laws of the *Koran* from the Faqihs and faithfully carry them out. Under these circumstances the real rulers are the Faqihs. Therefore, the government must officially belong to and be ruled by the Faqihs, and not by those who do not know the laws and have to follow the Faqihs."

Khomeini made no secret of his vision of an Islamic government, but it is obvious that no one paid attention. Those who should have known him better were, like the illiterate masses of Iran, caught up in the euphoria of the phenomenal Ayatollah.

In order to appreciate the powerful role of the Islamic Shi'ite in the realization of the goals of the Islamic Republic of Iran, it is necessary to consider the development of the Shi'ite sect through history up to the various political groups which exist in Iran today, using the label of Islam to pursue their various goals, as Khomeini did.

The Shi'ite sect of Islam would have remained much as the other minor sects (Jafaris, Hanafis, Ismaelis)–unknown and politically insignificant–had not the Arabs invaded Iran and converted the population to Islam in 642-651. At that time, the Shi'ite banner was a convenient one for the Iranians to use so that they could keep Islam but fight the Arabs, who were Sunnis. The Arabs were driven out of Islamic–now Shi'ite–Iran, and Iran continued to practice Islam, promoting the Shi'ite doctrine throughout the country.

The really big revival of Shi'ite doctrine came during the Safavid dynasty, beginning in 1499. For political reasons, the founder of the Safavid dynasty, Shah Ismail, created a false genealogy for himself which linked him to Imam Musa Kazim, the sixth Imam of the Shi'ites. Shah Ismail not only succeeded in uniting Iran under his rule, he also decided to convert the Sunni majority to the Shi'ite fold, by force if necessary. The major barrier to implementing this policy was the lack of Shi'ite scholars in Iran. The Safavids solved this problem by importing scholars from the Arab countries to help them with their undertaking.[83]

Since the revitalization of the Shi'ite sect by the Safavids, and its propagation to the extent that over 90 percent of Iranians (and over 75 percent of the population of the Persian Gulf region) are Shi'ites today, the Shi'ites have been influenced by four other ideologies: Marxism, Liberalism, Revolutionary Islam, and Progressive Islam.[84]

The influence of Marxism on Islam led to the creation, in October, 1941, of the Tudeh (Communist) Party of Iran, whose treachery during the Shah's reign has already been cited. After Khomeini's rise to power the Tudeh Party, instead of adopting a policy of confrontation with the Islamic Republic, he took a hypocritical stance of appearing to support the new regime, while devoting itself, with Soviet guidance, to creating a Party infrastructure throughout Iran. The end of the Tudeh Party came on February 5, 1983, when Nooruddin Kianoori[85], the general secretary of the Tudeh Party, was arrested together with other senior party leaders. Shortly thereafter, all the leaders of the Tudeh Party appeared on national television and confessed their acts of treason against Iran.[86]

The influence of liberalism on the Shi'ite sect brought with it the westernization of the Iranian culture. The liberals believed that in order to create a new culture in Iran, Islam must be set aside so that the modernization of Iran in the style of the West could be achieved. The pioneers of this ideology were Taghi-Zadeh, Sadegh Hedayat and Ahmed Kasravi.[87]

The reactionary Shi'ites were led by Ayatollahs Borujerdi, Khomeini, Kashani, and Hojat-ul-Islam Falsafi. The reactionaries sought to prevent the influences of liberalism and socialism on Islam. The most radical religious-political group, the Fedayane Islam in Iran, was also an outgrowth of the reactionary movement. With the victory of Khomeini's revolution, reactionary Islam experienced a renaissance.

The progressive Muslims were the fourth group that wanted to establish a constitutional Islamic government. At the political-reformist level, this Islamic development could be seen in the Liberation Movement founded by "moderate" Muslims like Mehdi Bazarghan, the first–but unsuccessful– prime minister of the Islamic Republic of Iran. At the revolutionary level, it could be seen in the progressive Islamic ideology being developed by Ayatollah Taleghani, Dr. Ali Shariati, and the Mujahidin-e-Khalq.[88]

The controversy surrounding Ayatollah Taleghani and Dr. Ali Shariati, as well as the political goals of the Mujahidin-e-Khalq, are issues which require some explanation. According to a publication of the Mouvement Iran Libre in Paris[89], the Soviets brought the Mujahidin into existence under the leadership of Ayatollah Taleghani. This was done in order to harness the power of Islam and its clergy to realize Soviet political goals in Iran. This is why the Mujahidin have been stamped "Islamic Marxists".

On the other hand, as soon as the British noticed the intentions of the Soviets and became aware of their resolve to penetrate the traditionally British domain of strategy (for example, working through the mullahs to attain political objectives) the British instantly installed Dr. Ali Shariati as a counterbalance to Taleghani. Thus, the Mujahidin ended up with two spiritual leaders.

Whether this political intrigue corresponds with reality is not relevant; what matters is the goals that these entities pursued. There is no doubt that Ayatollah Taleghani leaned somewhat to the left, and that Dr. Shariati and the Mujahidin-e-Khalq followed. Their paramount objective was to topple the Shah, do away with his regime, and obliterate the Pahlavi dynasty altogether.[90] Once this objective was achieved, the driving force for further action was subdued. Although Dr. Shariati was found dead on June 19, 1977, and he did not live to see the success of his actions, everywhere in Iran the demonstrators carried his portrait in parades.

Ayatollah Taleghani's fate was similar to that of his counterpart. He was released from prison in October, 1978, and soon became the nucleus of the revolution. However, his early death in September, 1979, removed one of the main hurdles in Khomeini's path, enabling Khomeini to consolidate his dictatorship in the name of religion.

The Mujahidin-e-Khalq went into decline after the revolution. Following a short honeymoon with the mullahs of the Khomeini regimes, they were outmaneuvered by Khomeini's strategists and pushed from the Iranian political scene. The mutual terrorism that followed this split drove the two factions farther apart. Hundreds of officials of the Khomeini government were killed. In retaliation, Khomeini had huge masses of young Mujahidin executed. The leaders of the Mujahidin escaped to France from Iran. Today they live in Iraq. Although the young Mujahidin sacrificed their lives for their ideology, their leaders' moral character and leadership qualities could both be described as weak.

Consider, for example, the Mujahidin leader Masud Rajavi. Rajavi's wife was killed by the Islamic Revolutionary

Guard in Iran, and their baby taken away. Rajavi escaped to Paris with Abolhassan Bani-Sadr, the first president of the Islamic Republic of Iran. There he married Bani-Sadr's daughter Mariam. A few years later, because of political differences with his father-in-law, he divorced his wife. Shortly thereafter, in a rare show of solidarity, a senior Mujahidin divorced his wife so that Rajavi could marry her and have the succor of a woman whose political bent was acceptable. According to the Mujahidin, this marriage took place with sacrifices on both sides, in order to strengthen the Mujahidin.

The Iranian people are skeptical of the Mujahidin-e-Khalq, as another Islamic political group which is, in addition to everything else, pro-Soviet.

Upon reflection on the ideologies of Ayatollah Taleghani and Dr. Ali Shariati, it appears that both men were devoted to the revival of the customs and laws of 1,300 years ago, and sought to tie them to the already failed economic and political doctrine of Marxism. Both men's ideologies were devoid of constructive teachings or principles, and neither attempted to guide his followers to improve their lot in life through education. In place of any plan for the betterment of their countrymen, they incited their followers to become revolutionaries and to topple the Shah, without realizing that the misery of Iran was not entirely the fault of the Shah. Thus, the intellect and energy of two brilliant men were wasted in the pursuit of destructive goals. With them perished a golden opportunity for constructive change, and the lives of count-less idealistic young Mujahidin.

An interesting note is that the objectives and philosophies of Islam and Marxism are irreconcilable. Islam promotes individual ownership of property and accumulation of wealth, and permits slavery. In the *Koran*, the class structure of society is endorsed, and the poor are told that their poverty is God-given and therefore their destiny.[91] The incompatibility of Islam and Marxism doomed the Mujahidin-e-Khalq to failure.

7

IRAN AND OPEC

S hould the Islamic Republic of Iran succeed in spreading its Islamic revolution throughout the countries of the Persian Gulf region, or join forces with the other fundamentalist movements, there is no question that the Islamic Republic of Iran would be in control of OPEC. The Islamic Republic would, after all, control over 76 percent of the world's oil resources. Suddenly the control of oil to the world would be in the hands of the fanatical mullahs.

The potential consequences of this situation are not difficult to imagine. There could easily be a revival of the unpleasant experiences of the West during the oil crisis of the early 70s and 90s. Other events of even greater economic and political impact could come to pass. Whatever might happen, the effects would be felt even more quickly than they were in the 70s.

The uncertainty in this prediction is due to the fact that the Islamic Republic would have a broad spectrum of options which it never had before. The Islamic Republic would have a free hand in wielding almost unlimited power, and the oil-importing nations would be powerless to take any effective countermeasures. Iran would have carte blanche to impose its will upon the world, because every nation of the world would either be directly dependent on the oil, or indirectly affected by the Islamic Republic's new financial power or change of monetary policy.

To understand the gravity of the circumstances which would face the oil-importing countries, it is important to look at some facts and figures. The Islamic Republic's domination of the world economy would revolve around oil, because oil is the primary source of energy in the world.

A comparison of oil reserves, oil production and oil consumption is shown in Table 2. The breakdown of the oil reserves, oil production and oil consumption of the Western industrialized nations is shown in Table 3.

Table 2

World Crude Oil Reserves, Production and Consumption in 1988[92]						
REGION	RESERVES MILLION TONS	%	PRODUCTION MILLION TONS	%	CONSUMPTION MILLION TONS	%
Middle East	88,855	66.1	712.5	23.9	146.0	4.8
Central & South America	16,827	12.5	337.1	11.3	236.7	7.8
Eastern Bloc Countries*	11,435	8.5	778.2	26.1	667.0	22.1
Africa	7,579	5.7	246.3	8.3	90.2	3.0
North America	4,486	3.3	548.0	18.4	863.0	28.6
Western Europe	2,455	1.8	195.8	6.6	562.0	18.6
All Others	2,846	2.1	160.8	5.4	455.9	15.1
WORLD TOTAL	134,483	100.0	2978.7	100.0	3,020.8	100.0
OF WHICH OPEC	103,041	76.6	984.6	33.1	162.7	5.4
Western Industrialized Countries*	6,941	5.2	743.8	25.0	1,640.0	54.3

* Eastern Europe, Soviet Union, People's Republic of China.
** North America, Western Europe, Japan.
Source: *Oeldorado 88*, Published by ESSO AG, Hamburg, 1989.

Table 3

Crude Oil Reserves, Production and Consumption of the Western Industrialized Countries in 1988[93]						
Western Industrialized Country	**RESERVES** Million Tons	%	**PRODUCTION** Million Tons	%	**CONSUMPTION** Million Tons	%
NORTH AMERICA	4,486	64.6	548.0	73.7	863.0	52.6
United States	3,572	51.5	455.0	61.2	787.0	48.0
Canada	914	13.1	93.0	12.5	76.0	4.6
WESTERN EUROPE	2,455	35.5	195.8	26.3	562.0	34.3
Great Britain	691	10.0	115.0	15.5	77.6	4.7
Norway	1,402	20.2	56.0	7.5	23.8*	1.5
Italy	108	1.6	4.5	0.6	86.8	5.3
Netherlands	59	0.8	4.3	0.6	32.9	2.0
West Germany	33	0.5	3.9	0.5	114.8	7.0
France	28	0.4	3.5	0.4	84.0	5.1
Others	134	1.9	8.6	1.2	142.2	8.7
JAPAN	—	—	—	—	215.0	13.1
TOTALS	6,941	100.0	743.8	100.0	1,640	100.0

* Includes Belgium, Luxembourg and Denmark.
Source: Oeldorado 88

These tables show that, while there is a balance between oil production and oil consumption overall, there is a major gap between production and consumption of the industrialized nations of the West, except for Great Britain and Norway, who are net exporters.

In 1988, the United States produced 455 million tons of crude oil, but consumed 787 million tons. 332 million tons, or 42 percent of consumption, was imported. The Western European countries were even worse off. In the same year, they had to import 366.2 million tons, or 65.2 percent of their total consumption. The Japanese were in the worst position; they had to import almost all of the 215 million tons they consumed in 1988.

In other words, with a total consumption of 1.64 billion

tons of crude oil in 1988, the industrialized nations produced only 743.8 million tons, leaving a shortfall of 896.2 million tons which had to be imported. These countries were dependent on imports for 54.6 percent of their oil consumption.

To fill the gap between production and consumption, the West can import from three different groups of oil-producing countries: members of OPEC, Eastern Bloc nations, and non-OPEC oil-producers. The reserves, production and consumption of these three groups are shown in Table 4. Their surplus available for export is analyzed in Table 4a which shows that 84 percent of the surplus oil available for export comes from OPEC countries, while only 16 percent is available from non-OPEC producers and the Eastern Bloc.

Table 4

Crude Oil Reserves, Production and Consumption of the Countries with Surplus Oil for Export in 1988[94]						
REGION	RESERVES		PRODUCTION		CONSUMPTION	
	Million Tons	%	Million Tons	%	Million Tons	%
OPEC MEMBERS	103,041	82.9	984.6	47.3	162.7	14.7
EASTERN BLOC	11,435	9.2	778.2	37.4	667.0	60.4
Soviet Union	7,959	6.4	924.0	30.0	450.0	40.7
People's Republic of China	3,226	2.6	135.0	6.5	106.0	9.6
CENTRAL & SOUTH AMERICA*	8,515	6.8	228.3	11.0	210.1	19.0
Mexico	7,362	5.9	143.0	6.9	77.4	7.0
AFRICA*	1,340	1.1	89.0	4.3	64.6	5.9
Egypt	594	0.4	44.5	2.1	21.5	1.9
Angola	281	0.2	22.5	1.1		
TOTALS	124,331	100.0	2,080.1	100.1	1,140.4	100.0

*Excluding OPEC members.
Source: Oeldorado 88

Table 4a

Oil Surplus Available for Export, 1988		
REGION	MILLION TONS	%
OPEC MEMBERS	821.9	84
Eastern Bloc	111.2	11
Africa*	24.2	3
Central & South America	18.2	2
TOTAL SURPLUS	975.7	100
*Excluding OPEC members.		

Table 5 analyzes the oil production shortfalls of the West. The extent of the West's dependency on imported oil is sobering.

Table 5

Gap Between the Production and Consumption of Crude Oil in the Western Industrialized Countries, 1988[95]						
WESTERN INDUSTRIALIZED COUNTRIES	PRODUCTION MILLION TONS	%	CONSUMPTION MILLION TONS	%	DEFICIT MILLION TONS	%
Western Europe	195.8	26.3	562.0	34.3	366.2	40.9
North America	548.0	73.7	863.0	52.6	315.0	35.1
United States*	455.0	61.2	787.0	48.0		
Japan			215.0	13.1	215.0	24.0
TOTALS	743.8	100.0	1,649.0	100.0	896.2	100.0
*The US deficit exceeds that of North America because Canada is a net exporter. Source: Oeldorado 88						

Table 6 shows the quantities of oil imported by the countries of the West from the member countries of OPEC.

Table 6

Crude Oil Imports of the Western Industrialized Countries from the OPEC Countries in 1987[96] (in million tons)

Western Industrialized Countries	Algeria	Ecuador	Gabon	Indonesia	Iran	Iraq	Kuwait	Libya	Nigeria	Qatar	Saudi Arabia	United Arab Emirates	Venezualla	Total
Western Europe	6.5	—	4.1	0.1	38.3	54.3	9.9	35.8	22.4	1.9	39.4	9.3	8.2	230.2
North America	0.9	1.5	2.7	9.2	14.0	4.9	2.1	—	27.2	—	31.6	4.5	25.7	124.3
USA	0.9	1.5	2.7	9.2	12.6	4.4	2.1	a	26.2	—	30.8	4.3	NA	94.7
Japan	—	—	—	17.8	11.6	5.2	7.8	—	—	.6	28.3	2.8	NA	106.1
TOTAL	7.4	1.5	6.8	27.1	63.9	64.4	19.8	35.8	49.6	7.5	99.3	43.6	33.9	460.6

Source: OPEC Annual Statistical Bulletin 1987, Published by the Secretariat Organization of the Petroleum Exporting Countries, Vienna, Austria, 1988, page 78-87.

Table 7 shows that of the total 896.2 million tons of crude imported by the Western industrialized nations in 1988, 460.6 million tons (51.4 percent) came from the OPEC countries, and 435.6 million tons (48.6 percent) came from non-OPEC countries. The shares of the Western countries' deficits which were covered by OPEC and non-OPEC countries are shown in Table 7.

Table 7

Shares of Oil Deficits of Western Industrialized Countires Covered by OPEC and Non-OPEC Countires in 1988[97]						
Western Industrialized Countires	Oil Deficits Million % Tons		Imports from OPEC* Million % Tons		Imports (Non-OPEC) Million % Tons	
Western Europe	366.2	40.9	230.2	50.0	136.0	31.2
North America	315.0	35.1	124.3	27.0	190.7	43.8
United States	322.0	—	94.7	20.6	237.3	54.5
Japan	215.0	24.0	106.1	23.0	108.9	25.0
Totals	896.2	100.0	460.6	100.0	435.6	100.0

*1987 figures
Source: Oeldorado 88 and OPEC Annual Statistical Bulletin.

Although there is sufficient production among the non-OPEC countries to make it possible for one or another of the oil-importing countries to switch to an all-non-OPEC supply for its imports, with all of the non-OPEC supplies capable of satisfying only 48.6 percent of the demand, the limited usefulness of switching suppliers is obvious.

The West's strong dependency on OPEC oil was the reason that OPEC was able to impose the price increases of the 70s and that the West was impotent to combat these measures. In the 1980s, however, due to the Iran-Iraq war and the need for income which pushed up production in the OPEC and non-OPEC oil-producing countries, the price of oil dropped significantly. This was exacerbated by the policies of OPEC, as largely dictated by the Saudis and Kuwaitis, who wanted to limit Iran's income and thus its ability to wage war. The beneficiaries of these polices were the oil-importing countries.

The Organization of Petroleum Exporting Countries (OPEC) was founded in September, 1960, in response to the unilateral decision of the multinational oil companies to cut posted prices in February, 1959, and again in August, 1960, thus inflicting severe economic damage on the economies and developmental programs of the oil-producing nations. The founding members of OPEC were Iran, Iraq, Kuwait, Saudi Arabia and Venezuela. Other oil-exporting countries were added, so that by 1973 membership stood at thirteen. The added members were Algeria, Ecuador, Gabon, Indonesia, Libya, Nigeria, Qatar, and the United Arab Emirates.

OPEC first began to gain momentum and recognition in 1971, when the Conference of Tehran set the stage for a permanent oil price increase during the 70s. The objectives of OPEC were spelled out clearly in its bylaws. Article Two reads:

A.   The principal aim of the Organization shall be the coordination and unification of the petroleum policies of Member Countries and the determination of the best means for safeguarding their interests, individually and collectively.

B.   The Organization shall devise ways and means of ensuring the stabilization of prices in the international oil markets with a view to eliminating harmful and unnecessary fluctuations.

C. Due regard shall be given at all times to the interests of the producing nations and to the necessity of securing a steady income to the producing nations; an efficient, economic and regular supply of petroleum to the consuming nations; and a fair return on their capital to those investing in the petroleum industry.

Table 8 shows the relative size of each OPEC member's crude oil reserves, production, and consumption. Iran occupies the fifth position among member countries, considerably behind Saudi Arabia, Kuwait, Iraq and the United Arab Emirates. Iran has 12.3 percent of OPEC's proven oil reserves, represents 11.5 percent of OPEC production, and only 9.8 percent (80.6 million tons) of OPEC's exports.

Table 8

Crude Oil Reserves, Production, and Consumption of the OPEC Member Countries in 1988[98]						
OPEC MEMBER	RESERVES Million TONS	%	PRODUCTION Million TONS	%	CONSUMPTION Million TONS	%
Iran	12,694	12.3	113.0	11.5	32.4	19.9
Saudi Arabia	34,464	33.5	251.0	25.5	34.8	21.4
Kuwait	13,499	13.1	73.0	7.4	N A	N A
Iraq	13,417	13.0	128.0	13.0	8.4	5.2
United Arab Emirates	12,893	12.5	76.3	7.7	7.6	4.7
Qatar	417	0.4	15.2	1.5	—	—
Algeria	1,070	1.0	30.0	3.1	7.4	4.6
Gabon	99	0.1	8.8	0.9	—	—
Libya	2,899	2.8	48.5	4.9	8.0	4.9
Nigeria	2,171	2.1	70.0	7.1	10.2	6.2
Ecuador	188	0.2	15.8	1.6	4.5	2.7
Venezulla	8.124	7.9	93.0	9.5	22.1	13.6
Indonesia	1,106	1.1	62.0	6.3	27.3	16.8
OPEC TOTAL	103,041	100.0	984.6	100.0	162.7	100.0
Source: Oeldorado 88						

Should the Islamic Republic of Iran spread its Islamic revolution to encompass the Gulf region, the Islamic Republic would ultimately control the oil reserves of Saudi Arabia, Kuwait, Iraq, the United Arab Emirates, and Qatar. Iran could also control the oil production of Bahrain, Oman and Syria, although this potential is not included in the figures in Table 9.

Table 9 reflects the changes in OPEC under Iranian domination. The membership of OPEC would be reduced to eight entities, including the Iranian Bloc.

Table 9

Potential Crude Oil Reserves, Production, and Consumption of OPEC Under The Domination of Iran based on 1989 figures[99]						
OPEC Member	RESERVES		PRODUCTION		CONSUMPTION	
	Million Tons	%	Million Tons	%	Million Tons	%
IRANIAN BLOC*	87,384	84.8	656.5	66.6	83.2	51.2
Algeria	1,070	1.0	30.0	3.1	7.4	4.6
Gabon	99	0.1	8.8	0.9	—	—
Libya	2,899	2.8	48.5	4.9	8.0	4.9
Nigeria	2,171	2.1	70.0	7.1	10.2	6.2
Ecuador	188	0.2	15.8	1.6	4.5	2.7
Venezuella	8,124	7.9	93.0	9.5	22.1	13.6
Indonesia	1,106	1.1	62.0	6.3	27.3	16.8
OPEC TOTAL	103,041	100.0	984.6	100.0	162.7	100.0

*Includes Saudi Arabia, Kuwait, Iraq, the United Arab Emirates and Qatar.
Source: Oeldorado 88 and the author's calculations.

By controlling 84.8 percent of OPEC's crude oil reserves and 66.6 percent of OPEC's oil production, the Islamic Republic would be in an invincible position to impose its will on the oil-importing nations of the West. Furthermore, the Islamic Republic could easily assume control of the world's monetary system, by unleashing the power latent in control of a vast supply of oil by coupling it with a shrewd monetary policy. Iran could become the financial center of the world.

In the 70s OPEC created havoc in the industrialized countries by raising the price of oil. The same thing could happen again, but this time the options would all belong to one oil-producing country: the Islamic Republic of Iran. And Iran could establish this trend single-handedly, with all of the other oil-producing countries following closely behind.

Should this event come to pass, the oil-importing countries would try to retaliate, as they did in the 70s, by adopting a hostile policy toward Iran and OPEC, and clinging desperately to their limited options. The West would then try to purchase all the crude oil that they needed from non-OPEC suppliers, but this would provide only temporary relief. The countries that are not members of OPEC are referred to as NOPEC. NOPEC consists of Angola, the People's Republic of China, Columbia, Egypt, Malaysia, Mexico and Trinidad.

The strategic importance of the NOPEC oil surplus must be acknowledged. There is no question that it could provide some respite, however temporary. But one does not need to be a brilliant mathematician to see that a shift to NOPEC oil could not conceivably replace the flow of OPEC oil. Nor could one imagine that such a shift could neutralize Iran's control of OPEC, or Iran's economic or political decisions.

The very most that could be expected of the NOPEC countries would be for them to supply the requirements of one or two of the industrialized nations on any sort of permanent basis. This could, of course, be of some strategic benefit to the fortunate one or two countries, but as for the rest of the world, there are the limitations of NOPEC oil. With the exception of Mexico, the NOPEC countries are insignificant producers of oil compared to their OPEC rivals.

Table 10 compares the oil reserves, production and consumption of the OPEC and NOPEC countries. The relative insignificance of NOPEC in a global context is clear.

Table 10

Crude Oil Reserves, Production, and Consumption of OPEC Under Iran and NOPEC Countries, based on 1988 Figures[100]						
	RESERVES Million Tons	%	**PRODUCTION** Million Tons	%	**CONSUMPTION** Million Tons	%
OPEC	103,041	89.4	984.6	71.3	162.7	41.7
Iranian Bloc	87,384	75.8	656.5	47.5	83.2	21.3
Other Members	15,657	13.6	328.1	23.8	79.5	20.4
NOPEC	12,205	10.6	397.2	28.7	227.7	58.3
Angola	281	0.2	22.5	1.6	—	—
People's Republic of China	3,226	2.8	135.0	9.8	106.0	27.1
Colombia	287	0.3	17.4	1.3	8.1	2.1
Egypt	594	0.5	44.5	3.2	21.5	5.5
Malaysia	379	0.3	27.0	1.9	10.2	2.6
Mexico	7,362	6.4	143.0	10.3	77.4	19.8
Trinidad	76	0.1	7.8	0.6	4.5	1.2
TOTAL OPEC + NOPEC	115,246	100.0	1,381.8	100.0	390.4	100.0

Source: Oeldorado 88 and the author's own calculations.

Table 10 demonstrates that with OPEC under its control, the Islamic Republic of Iran would control over 75.8 percent of the total reserves and over 47.5 percent of the total oil production of OPEC and NOPEC combined. The surplus of 573.3 million tons of crude oil would be equal to more than the total production of the United States and Canada. Furthermore, as Table 11 shows, the Iranian Bloc's surplus oil for export would be nearly one and one-half times greater than the combined surplus of the other OPEC members and that of the NOPEC countries. Almost 58 percent of OPEC and NOPEC oil for export would come from the Iranian Bloc, making the Iranian Bloc the largest exporter of oil in the history of the world.

Table 11

Potential Surplus Oil for Export from OPEC under Iran and NOPEC Countries, based on 1988 figures[101]		
	SURPLUS	
	Million Tons	%
OPEC	821.9	82.9
Iranian Bloc	573.3	57.8
other OPEC Members	248.6	25.1
NOPEC	169.5	17.1
Angola	22.5	2.3
People's Republic of China	29.0	2.9
Colombia	9.3	0.9
Egypt	23.0	2.3
Malaysia	16.8	1.8
Mexico	65.6	6.6
Trinidad	3.3	0.3
TOTAL OPEC + NOPEC	991.4	100.0
Source: Oeldorado 88 and the author's own calculations.		

One cannot overlook the fact that it is unlikely that the NOPEC countries would sacrifice their own advantages for the interests of the industrialized countries. If it were not for the severe threats facing Saudi Arabia, Kuwait and the United Arab Emirates due to the war between Iran and Iraq, oil prices would have been subject to increase. Even with the political and economic implications of increased oil prices, there is no doubt that the oil-exporting countries would end up net beneficiaries of higher oil prices, in a very big way.

The oil price increases of the 70s brought economic boom not only to the OPEC countries, but heaped incredible bounties upon non-OPEC countries, such as Mexico. Even Great Britain enjoyed higher income from its exploitation of the North Sea. The multinational oil companies registered record incomes and profits, and the economy of Texas boomed as never before. Therefore it was no accident that the officials of the NOPEC countries met with their OPEC counterparts at OPEC's May, 1988, conference. Worthy of note was the presence at this conference of a former three-term United States Congressman, Kent Hance, a member of the Texas Railroad Commission, which regulates the production of oil in that state.[102] Each and every participant in the oil business benefits from rising oil prices; to imagine otherwise would be illogical and naive.

The conservation of energy and more rational use of oil could reduce the dependence of the West on imported oil. The effectiveness of such measures to date has been insignificant. Even though the consumption of oil showed a decline from 1980, the recent trend is toward more consumption, as shown in Table 12.

Another possible means of reducing dependency on imported oil would be for the importing nations to initiate and expand development of alternate energy sources, such as solar, nuclear, tar sand, shale and coal, among others. Before too many hopes are pinned on such developments, it would be wise to consider them in the context of reality.[103]

During the 70s it was generally assumed that an increase in the price of crude oil would automatically stimulate the development of alternate energy sources. The economic ra-

Table 12

Crude Oil Consumption Worldwide 1975, 1980,1985-1988 (Millions of Tons) [104]						
	1975	**1980**	**1985**	**1986**	**1987**	**1988**
NORTH AMERICA	846.9	889.5	792.6	823.1	842.0	863.0
U.S.A.	759.8	799.8	723.3	752.9	767.9	787.0
WESTERNEUROPE	642.9	648.9	542.0	562.1	563.8	562.0
West Germany	128.8	130.5	112.9	119.9	115.2	114.8
France	107.6	110.5	82.3	84.0	84.5	84.0
SOUTH & EAST ASIA &AUSTRALIA	391.2	456.8	431.4	441.5	444.4	455.9
Japan	243.3	260.4	206.9	213.5	207.9	215.0
EAST BLOC, U.S.S.R. & PEOPLE'S REPUBLIC OF CHINA	536.3	655.4	635.8	646.0	665.0	667.0
CENTRAL & SOUTH AMERICA	184.8	223.5	227.1	224.8	233.7	236.7
AFRICA	55.4	72.6	89.3	89.9	89.4	90.2
MIDDLE EAST	75.4	106.5	141.2	144.9	145.0	146.0
WORLD TOTAL	2,732.9	3,062.2	2,859.4	2,932.3	2,932.3	3,020.8

Source: Oeldorado 88

tionale upon which this assumption is based submits that the higher the price of energy each year, the more the consumer will lean towards more frugal use of energy. At the same time, the potential for profits will be so attractive as to stimulate the producers toward greater investment in alternate energy sources.

The simultaneous occurrence of reduced energy consumption and expansion of energy supplies should enable domestic energy production to grow to the extent of near-self-sufficiency, according to the theory. Simply put, the rise of the price of oil will inevitably lead to measures being taken to find a substitute for oil, or so it has been assumed.

Rising oil prices resulting in substitutes for oil may be true in microeconomic terms; however, in macroeconomic terms the validity of this assumption is only maintained to a certain extent. The assumption that rising oil prices will accelerate the development of substitute energy sources is not realistic in the context of a fully functioning market economy, which must serve as a basic assumption. Should crude oil prices increase, normally the demand for crude oil will sink, at least as long as the possibility of substitute energy sources exists. Decreasing prices of crude oil resulting from lower demand also influence the substitution process, while energy consumption begins to rise again, so that in the long run oil prices will fluctuate cyclically with an upward trend.

The extent to which the substitution for crude oil is feasible rests on answers to several questions: When will the substitution be possible? Will the resources necessary to replace crude oil be adequate in quantity as well as quality? What will be the prices of the substitutes? To what extent will the use of the substitutes be detrimental to the environment, and what will the eventual social costs be?

It must be remembered from the outset that an efficient substitute for crude oil can only be assured if the substitute is abundantly available and meets quality requirements. The closely interrelated economic, technological and ecological problems must be solved. A favorable compromise of solutions will take time, especially since there are many uncertainties about the interdependencies of the relevant problems. Possible intensification of alternate energy development depends on the extent of research efforts, which in turn depend on economic factors such as insufficient investment or conflicting targets in economic policy. Attempts to accelerate the substitution process by employing

technologies which are not fully developed may cause difficulties later on.

For example, if nuclear energy should be chosen to substitute for part of a nation's oil consumption, the building of
conventional nuclear reactors must be accelerated. However,
during the more than ten years required to bring a nuclear
reactor on-line, the demand for its output (electricity)
generally outstrips the new source of supply. Meanwhile, in
the ten years of construction, the development of new technology makes the plant obsolete before it is finished. And, of
course, the growing demand for electricity makes it mandatory to build new reactors faster than ever. Large sums of
money are wasted on applications of technology which is not
completely developed.[105]

The social cost and environmental impact of nuclear
power are well known, if debatable.

In any event, the development of viable alternate energy
sources will require time, even assuming that the necessary
financial resources are available. The time required will also
be determined by the exigencies of putting the new technology into production on a scale that will provide meaningful
quantities of energy.

In considering the feasibility of substitutions for crude oil,
it is important to distinguish between quantitative and
qualitative substitution. The qualitative aspect of crude oil,
for the sake of this discussion, is that oil provides not only
fuels, but fibers, plastics and fertilizers as well. Nuclear
power, for example, can substitute for oil only in generating
electricity or heat. On the other hand, tar sands, shale, and
coal could substitute for oil qualitatively in many ways.

Although a transformation of organic substances into synthetic oil and gas may be achieved by the application of

electrical energy, and electrical energy can also transform these substances into other products which substitute for oil products, the electrical energy still has to come from somewhere, and the process is thus quite inefficient.

Rising oil prices, then, cause only limited amounts of substitution for crude oil. To be significant, oil substitutes must be available in vast quantities, at prices comparable to that of crude oil. The chart illustrates the possible situations concerning substitution for crude oil.

Further, it must be noted that if a country uses more costly energy derived from domestic resources to reduce dependency on imported oil, trade deficits from the resulting deteriorating international trade position would be just as bad as those caused by importing oil.

Rapid development of alternative energy sources to the extent that some modicum of self-sufficiency is achieved requires massive capital investment. In order to begin to envision the magnitude of investment necessary, consider

Description of Possible Energy Substitution Situations[106]		
SUBSTITUTIVE ENERGY SOURCE		DESCRIPTION OF SUBSTITUTION PROGRESS
Volume	Price	
Sufficient	Higher than oil	No substitution: oil prices may rise until they reach the price of the substitutes.
Insufficient	Higher than oil	No substitution: oil prices may rise further.
Insufficient	Lower than oil	No substitution: The prices of substitute energy approach oil prices.
Sufficient	Lower than or equal to oil.	Total substitution for crude oil or price of crude oil falls.

that official estimates of the capital required to accomplish the "failed" Project Independence[107] in the United States in the 70s ranged from 450 to 700 billion dollars; the Apollo Project cost only 225 billion.[108]

The investment required to achieve independence from imported oil raises the problem of financing the venture. Furthermore, the intertemporal distribution of the burden is an issue requiring consideration. From a social point of view, it seems unfair to burden a generation which has neither gained from times of cheap energy nor will benefit from the investment in the future. In economic policy, such considerations might lead to conflicting objectives which are aggravated by existing economic relationships. These incompatibilities must be included in economic assessments.

According to one study, "if the upset in international markets leads only to a worldwide two-year zero-growth recession, it will be as costly as four years of extra oil payments."[109] Thus, rising oil prices are only one of a number of economic factors which must be weighed when contemplating the development of alternate energy sources.

The ecological side-effects of conventional energy production are well known. However, some of the technologies contemplated or under development pose their own environmental problems. Solar energy, known to be the cleanest energy source, may prove to have an unpleasant environmental impact if it is ever realized on a large scale. Generation of electricity on solar farms would require 14,000 square miles of solar collectors to accommodate America's electrical consumption until the year 2000. This is roughly equal to the combined areas of Connecticut and Massachusetts.

If solar energy were produced by the Satellite Solar Power System, the microwaves beaming the energy from space would be potentially hazardous. On the other hand, there are no environmental hazards from solar homes or simple solar heating and cooling systems.

Waste from nuclear power plants and carbon dioxide from burning coal are but two of many examples of environmental incompatibility of alternate energy sources which will limit their development. To quantify the environmental incompatibility of energy sources, it is necessary to distinguish between primary and secondary energy sources. Secondary sources have little impact upon the environment; most primary sources of energy present problems. For example, the electricity (secondary source) generated by a nuclear power plant is a clean form of energy. The nuclear power plant (primary source: uranium) generates hazardous waste. The production of synthetic oil and gas from coal, tar sand, shale or waste creates environmental pollution. However, the electricity generated from synthetic oil or gas would be a clean secondary energy source.

This foregoing analysis demonstrates the acute worldwide dependency on oil, and that the likelihood of efficient substitution for oil of alternate energy sources is limited to none in the foreseeable future. Therefore, it is no wonder that today, at the beginning of the 1990s, the world is at least as dependent on crude oil as it was in the 70s. The Islamic Republic of Iran, as the potential dominant force in OPEC today, could have carte blanche to dictate economic and political policy to the world.

8

OPPORTUNITY FOR SUPREMACY

*H*ad Khomeini the wisdom to accept the peace desperately offered by the Iraqis, Saudis and other Arab nations in 1982, or had the Islamic Republic won the war with Iraq, Iran could easily have assumed the lead position in the Middle East and in OPEC, making itself a global superpower.

Supremacy for Iran could have been realized through the coupling of the control of the flow of oil with a new monetary policy, which would not have been difficult for the Islamic Republic to impose upon the oil-importing countries worldwide. The implementation of this policy would have triggered a series of chain reactions, which would have created drastic changes in the economic and political scenery worldwide. One development of special significance would have been the dramatic weakening of the "hard" western currencies of today, with the U.S. dollar, the West German mark, the Japanese yen, the British pound and the Swiss franc all declining in the face of a new leading currency: the Iranian rial.

The rial would have undergone a permanent, radical upward revaluation against all other currencies. The rial, which has fluctuated from 70 to the U.S. dollar in the time of the

Shah, to a low of 1400 to the dollar at the height of the Iran-Iraq war, would have probably risen to a level of ten to twenty rials to the U.S. dollar. This would have placed the rial among the reserve currencies of the world.

Obviously, this is not what happened. The question is, does this threat still exist?

Even without the Islamic Republic gaining control in the Middle East and OPEC, such a policy could be initiated anyway, for example, by a unified OPEC. One can understand this by examining some of the common misconceptions which are common among the industrialized, oil-importing countries.

There are many reasons for Iran and OPEC to couple the flow of oil to a monetary policy, and not to repeat the mistakes of the 1970s, when the flow of oil was coupled only to price increases. The drastic increases in the price of oil did not bring a level of prosperity to the producing nations commensurate with the new price levels. During the 1970s, Iran and the other OPEC countries received devalued dollars, which were used to purchase imported industrial goods and military equipment at highly inflated prices. Many of the inflated dollars were wasted paying for imports that spoiled or rotted in OPEC ports due to the bottleneck caused by insufficient inland transportation. Even after this economic activity, a massive surplus of dollars from the flow of oil remained. These "petrodollars" found their way back to the banks in Great Britain, the United States, West Germany, France, and other countries, to be recycled.

The inhabitants of the oil exporting countries, especially those with lower populations and higher revenues, enjoyed major shopping sprees, snapping up automobiles, consumer goods of every description, and an occasional power plant or factory. They traveled abroad to purchase things unavailable to them at home, or were exorbitantly costly there. But this sudden wealth was superficial and temporary; permanent prosperity did not come to the OPEC countries in a way to benefit all its citizens. Although more money was available than ever before, a growing number of people became dissatisfied with the runaway inflation, bottlenecks and shortages–especially in Iran–to the point where the foundation of the revolution was laid.

Furthermore, in the case of Iran, an immense unused portion of the oil windfall did not bring prosperity to the country. After the revolution, and later during the hostage crisis, the United States confiscated over $15 billion of Iranian deposits in U.S. banks. This event reinforces the concept that the currency of a country is only as good as that country's intent to honor it, especially if the currency is on deposit in the issuing country.

To clarify the problems associated with the increase of oil prices, one must begin where the problems begin, with the pricing of oil itself. The price of crude oil could be raised to any desired level before a substitute for crude oil could be put in place–at least a period of several years. On the other hand, an uncontrolled rise of the price of oil is neither politically nor economically advisable. Consequently, an optimum pricing policy is required, which is not possible. In practice there are three things that can be done with oil prices: increase them, freeze them, or decrease them.

An increase in the price of oil, as suppliers try to maximize their income from oil, would produce the same effects mentioned earlier: runaway inflation, bottlenecks at the docks, shortages. A freeze or decrease in the price of oil cannot be considered at all, because either action would worsen the terms of trade for the oil-exporting countries. The experience of OPEC prior to 1971 underscores this fact, as a result of which there were dramatic increases in the price of oil in the 1970s. A logical oil price strategy will tend toward an increase in the price of oil.

An optimum pricing policy for oil requires consideration

of the following factors:[110]

- The possibilities and costs of substitutes for crude oil;
- The fact that oil reserves are finite and do not regenerate;
- Compensation for the potential loss of value of the U.S. dollar and the British pound, the official currencies of oil transactions today;
- The probable price increases of products imported by the oil-exporting countries, exacerbated by the increased price of oil;
- World economic and political interdependencies; and
- The economic development goals of the oil-exporting countries.

The difficulty in evaluating these factors is that there are conflicts between them. The first four factors strongly favor permanent price increases, while the fourth and fifth also point toward as low a price as possible, and the sixth is ambiguous, because the goals for economic development differ from country to country.

Permanent escalation of oil prices as a means to maximize the income of the exporting countries, and thereby optimize the benefits, simply cannot be achieved. In other words, it is not possible to determine exactly when and how much the price of oil should increase, in a way that would maximize the real gain of the exporters.

Since many of the oil exporting countries are members of a cartel (OPEC), as long as they can agree on policies it would not be difficult for them to increase the price of oil as much and as frequently as they please. This will be true as long as OPEC meets the criteria of a viable cartel.

In order to have a functional raw material cartel, the following essential prerequisites must be satisfied:

1)There must be low elasticity of demand for the commoaity. This means that the commodity must be consumed regularly, with demand not affected significantly either by conservation, or by substitution of a cost-effective alternate commodity. In this situation, price increases would therefore have little impact on demand.

2)The cartel must control a sufficient portion of the supply of the indispensable commodity to be assured that the cost to the consumer of dispensing with the commodity will be greater than the harm caused the producers by any loss of sales.

3) The commodity should not be perishable or expensive to store, to reduce pressure for compulsory sales.

The prerequisites of a powerful cartel, with potential for long-term endurance, are optimally fulfilled by OPEC in the oligopolistic world oil market. However, without carefully coordinated action on the part of the members, the cartel's position could be at risk. This is because crude oil is more or less the same wherever it comes from. If one member of the cartel were to reduce the price only slightly, theoretically all the buyers would switch to that source to the extent that it could supply. In this situation, the cartel would collapse.

Throughout the 1970s, OPEC managed to prevent such a situation. However, when war between Iraq and Iran began, OPEC fell into disarray and seemed in danger of collapse in the 1980s. Once the disputes within OPEC are settled after the war, it is very possible for OPEC to regain is unity and begin to function again. The potential dominance of the Islamic Republic of Iran in the cartel could well influence the other members to follow Iran's policies and to steer the political course of Iran's choosing.

What kind of policy should the Islamic Republic pursue? And one must realize the shortcomings and limitations associated with a policy of permanent oil price increases. Whatever the policy, it must revolve around the following goals:

1)Optimization of economic benefits, especially by reducing the cost of imported goods for Iran, stabilization and increase of real income, and coordinated industrialization of the country;

2)Minimizing the threat to the position of crude oil which is latent in the development of alternate energy sources, which, in turn, is stimulated by high prices for oil; and

3)Minimizing the economic disadvantages of the oil-importing countries due to the secondary inflation process touched off by rising oil prices. Because oil is the primary energy and raw material source in industry and the world economy, any increase in the price of oil will immediately trigger an inflationary process. The increased cost of energy and materials is marked up appropriately and passed along to be marked up again at each stage from production through distribution and consumption of the affected goods and services. Since wages are indexed to the cost of living, wages are adjusted and wave after wave of secondary inflation results. This, of course, increases costs of imports to the oil-producing countries as well.

An ideal way for the Islamic Republic to achieve these goals is to couple the control of the flow of oil with a monetary policy which would allow Iran's currency, and possibly those of other OPEC members, to float and revalue their exchange rates. The Iranian rial would gain a position equivalent to that of the dollar, the mark, or the pound in the world economy. This policy could have some negative effects on Iran's economy, but, as will be pointed out, the advantages would greatly outweigh the disadvantages.

Replacing existing oil pricing policies with this new monetary policy coupled with control of the flow of oil would be simple and practical. The first step to implementation would be to convert the pricing of oil from dollars and pounds to Iranian rials. Consequently the other OPEC members would follow suit and receive value in their native currencies. In other words, oil would be priced in Iranian rials, and would have to be paid for in rials if one wanted to get oil from Iran. For the sake of this discussion, only the case of Iran will be considered, although the same could occur with other OPEC members. The extent of Iran's ability to implement this policy would be primarily dependent upon timing; the central and private banks of the oil-importing countries would have to have sufficient supplies of rials on hand to pay for the oil in that currency.

Every new economic or political situation requires a period of adjustment, so it would be necessary to begin the implementation of this policy by accepting payment partially in rials, with the balance in dollars, marks or pounds, until enough rials were in circulation to convert completely to rials.

One might suppose that Iran's desperate need for foreign currency to pay for its own imports would prevent implementation of such measures. This fact, however, would not provide the least impedance; the oil-importing countries would happily advance the required currencies to Iran, especially when the currency in question is the currency of the lender.

Next would come the change of exchange rates of the rial for other currencies. The Islamic Republic could revalue its currency in agreement with the International Monetary Fund, but this should be avoided. The importance of crude oil, not the IMF, should be the revaluing factor. The convertibility of the rial would not arise, or could be prevented, were it not for crude oil backing up the rial's new function. While the convertibility of a currency does not always bode well for it (it can be devalued), a revaluation of the currency can enhance profits. This is more likely in the case of the convertibility of the rial.

Then there is the matter of determining the price of a barrel of oil in rials. Although the easiest method would be to use the exchange rate of the dollar to the rial to figure the price, in so doing it would be very important to keep the exchange rate very low, to limit the quantity of rials in circulation. Efforts would have to be made to keep the rial in short supply. Should the rial establish itself as a leading currency, its limited availability would assure its convertibility, and would also allow Iran to manipulate the world of finance.

The first consequence of this economic/political measure would be the reaction of the major oil companies. In order to be able to make payments in rials, they would have no choice but to obtain this currency. In the early stages it would not be too difficult for the banks of the oil-importing countries to acquire enough rials. But, since the availability of rials would be limited, the process of revaluation and convertibility of the rial would begin. It would be incumbent upon Iran to follow a well-measured and tailored currency policy with a flexible exchange rate that would work to Iran's advantage.

Subsequent to the implementation of the new policy, goods imported by Iran would become less expensive. Today, Iran and other OPEC countries subsidize the prices of many imported consumer goods from oil revenues, to neutralize the effects of inflation and price increases. Under the new policy, a substantial amount of oil income would be saved, due to lower prices of imports and further due to eliminating the need for subsidies.

Since the oil-importing countries would desperately need rials, they would drastically increase their offerings of goods and services to obtain them; they would reduce prices in competition for the short supply of rials, which could only be obtained through the sale of goods and services. Additional motivation to improve their offerings would come from the oil-importing countries' expectations of profits from the anticipated revaluations of the rial. For Iran, this would mean reversal, prevention, or at least minimization, of price increases in the goods which it imports. Iran would receive an abundance of goods and services from competitive suppliers, and its worries about delivery bottlenecks would be over; since there would be great opportunities for suppliers to make money, it would be the responsibility of the suppliers to deliver their products to the market.

Foreign countries exporting to Iran would find great incentives for direct investments in Iran, because of the market potential and the potential gains from exchange rates. To clarify this point, consider a simple example, with the U.S. dollar equal to one Iranian toman (1 toman = 10 rials), and a product selling to Iran at $90, or 90 tomans. The product then sells in Iran for 100 tomans ($100). Should the exchange rate of the toman be revalued against the dollar by 10 percent (so that $1 = 90 rials or .90 toman), the same product selling to Iran for $90 would cost the importer only 81 tomans, and the importer gains by 9 tomans. If the product were produced in Iran, the manufacturer would benefit from the 9 tomans. A further gain could be realized by an American company operating in Iran when it transferred toman profits to the United States by converting them to dollars. Each 90 tomans would yield $99.00. This gain would be at the expense of the Federal Reserve Bank of the United States.[111]

This situation would benefit Iran by the importation of technology from the hypothetical American company, creation of new jobs for Iranians, and dampening the effects of imported inflation. The impact of Iran's coupling of control of the flow of oil with the new monetary policy would be felt around the world as follows:

IRAN

Iran's oil income and foreign currency reserves would not decline in purchasing power as they have in the past; the inflationary forces affecting the prices of Iran's imports would be virtually neutralized, as would potential losses from exchange rates. Since only a partial and indirect oil price increase would occur, the inflationary process would not become a certainty. However, a limited amount of inflation cannot be completely avoided, due to the *de facto* increase in the price of oil created by the revaluation of the rial against the other major currencies.

The effects of oil-price-induced inflation upon Iran's economy would be negligible, because they would be offset by the revaluation of the rial. At the same time, the development of alternate energy sources, which would have bee∴ stimulated by continually rising oil prices, would be suppressed.

Goods and services from oil-importing countries would become more and more competitive, and foreign investment in Iran would be greatly stimulated.

Due to its accumulation of wealth and the new significance of the rial, Iran would become a major financial center of the world. As a direct result of this, no foreign power would ever again dare to interfere with Iran's political destiny, nor meddle with Iran's assets abroad. Iran would be doubly protected from such threats. Would-be aggressors would have rials on deposit in Iran, so any hostile measures would boomerang. Secondly, with the rial as a reserve currency, no country would want to jeopardize its reserves; indeed, any holder of a currency has a vested interest in the prosperity of the currency's issuer. The stronger a country's economy, the stronger its currency.

Looking back at the 1970s, it is interesting that none of the OPEC countries took advantage of the opportunity to couple this sort of monetary policy with control of the flow of oil. It was an idea whose time had not yet come.[112]

THE INDUSTRIALIZED NATIONS

Although the implementation of the new monetary policy would not bring the same advantages to the oil-importing countries of the West, one positive aspect of these measures would be the dampening of inflation in the West, which otherwise would have been induced by rising oil prices.

THE DEVELOPING COUNTRIES WITHOUT OIL

The burden falling on this sector of the world's economy under the new monetary policy would be less than that imposed by constantly increasing oil prices. This amelioration would be due to the dampening of inflation which had been stimulated by rising oil prices.

The only significant disadvantage to Iran which could result from the implementation of the new monetary policy would be avoidable. While its exports of crude oil could not possibly suffer from careful application of these measures, exports of non-oil products could suffer if their prices were also fixed in rials. Although Iran's exports of non-oil products are presently minimal compared to its oil exports, as Iran prospered under the new monetary policy and industrialized to a greater extent, especially with the influx of foreign investment, the share of non-oil products would show rapid growth. If pricing these products should impede sales (after all, similar products are available elsewhere), then these products can simply be priced in dollars or other currencies.

Table 13

Iran's Exports of Oil and Non-oil Products (Millions of Dollars)				
YEAR	CRUDE OIL	NON-OIL	TOTAL PRODUCTS	SHARE OF NON-EXPORTS OIL PRODUCTS (%)
1970	2,358	45	2,403	1.9
1975	19,634	547	20,181	2.7
1980	13,286	820	14,106	5.8
1985	13,115	213	13,328	1.6
1986	7,199	1,139	8,322	13.7
1987	10,515	1,170	11,685	10.0
1988	8,170	914	9,084	10.1

Source: OPEC Annual Statistical Bulletin 1988.

Table 13 compares Iran's oil exports to its exports of non-oil products.

To maintain orderly growth in export sales of its non-oil products, Iran could also employ subsidies on certain products, or take advantage of manipulations of exchange rates.

Skeptics may question whether Iran could implement such a monetary policy in the existing world monetary system. After all, the world monetary system is based upon the currencies of the industrialized Western nations, which contribute to 67 percent of all world trade. The economic power of these nations is much greater than that of any other group of countries, and Iran is only a developing country.

This is all true. However, Iran could readily implement its polices, especially should the Islamic Republic take control of OPEC, which dominates the world's supply of crude oil, which is the key to the existence of the economies of the Western industrialized nations.

The fact that Iran could realize its economic goals autonomously should not be surprising. The experience of OPEC since October, 1973, is ample proof that economic policies concerning crude oil are unilaterally in the hands of the suppliers. The real question is how long it would take Iran to implement these policies should this country establish itself in a position of dominance in the Middle East and OPEC. It took OPEC nearly three years (February, 1971 until October, 1973) to realize that no bargaining was required when it came to the price of oil. The implementation of this policy could happen immediately; the only prerequisite is unity among the members of OPEC. Whether this were established through Iran's domination of the Middle East and OPEC, or otherwise, is secondary. More important is the threat that such a policy could be realized. Once its unlimited potential benefits were recognized, it is possible that this itself would create the necessary unity among the members of OPEC to implement the policy overnight.

9

THE LEGACY OF KHOMEINI

K homeini succeeded in ousting the Shah and forming the Islamic Republic of Iran, but after a decade in power he died on June 4, 1989, having failed to fully accomplish the goals described in his book *Hukumate Islami (Islamic Government)*.

Although Khomeini's goal was the worldwide domination of Shi'ite theology rather than economic and political stature such as that of post-war Germany and Japan, had Khomeini succeeded in spreading his Islamic revolution throughout the Middle East, Iran would have become a superpower as a matter of course.

Never in Iran's 3,000 year history was there a leader like Khomeini, who so moved the people, yet so callously abused them. And never before was there such a funeral, with millions of hysterical mourners turning out to bid Khomeini farewell. Many of the mourners were there voluntarily; many others came for food coupons, money, or under military threat.

In 1978, few would have predicted that Khomeini would take power in Iran just a year later. One wonders where

Khomeini would have ended up had the Shah chosen to resist the revolution with the military force he had available, or if the Shah had transferred power to one of his hard-line generals instead of to Shahpour Bakhtiar, his failed last prime minister.

Did Khomeini's Islamic political ambitions die with him, or will his goals ultimately be achieved? However speculative, the best answers will be found in the Islamic Republic's balance sheet for the last ten years, and in the backgrounds and aims of Khomeini's successors and of those who are struggling for power in Iran today.

What did not occur upon Khomeini's death was what many people predicted: a chaotic power struggle among the opposing factions in Iran. In fact, there was an unexpected show of unity among the leaders of the Khomeini regime, and a very smooth assumption of power by the first of Khomeini's successors. Considering the terror tactics practiced by the Khomeini government and its membership, it is hard to imagine that things are settled as peaceably as they seem to have been.

When Khomeini first came to power, he designated Ayatollah Hossein-Ali Montazeri to succeed him as supreme spiritual leader, and to interpret, guide, decide and generally rule the country according to Khomeini's theology, as a *Valiye Faghih.*

Khomeini chose Montazeri as his successor because Montazeri was a former student of Khomeini and loyally shared Khomeini's views.[113] Montazeri had other attractive attributes: he had been imprisoned by the SAVAK, which raised his stature as a political teacher after he returned to Qom—his classes were very popular among the young *talabehs* (theological students)[114]; further enhancing Montazeri's public image was the fact that he had been imprisoned with Ayatollah Taleghani before the revolution.[115] Khomeini considered Montazeri to be a simple man, who posed no threat.

There are two conflicting views of Montazeri. He is described by some as a very humble, simple and liberal man, who had no interest in material or political gain.[116] Others say the opposite, that Montazeri was a radical clergyman, shrewd and intelligent, who had given orders to many radical and terroristic organizations in Iran and abroad. Montazeri has been implicated as a leader of the "Organization for the Movement of Liberty", a terrorist group who kidnaped people in Iran and elsewhere after the revolution, and as a creator of the "Hezbollah" (Party of God) in Lebanon.[117] According to this view, Montazeri was involved in plans for hijacking airliners, which were executed by Mehdi Hashemi and his group.[118]

Mehdi Hashemi was one of Iran's most colorful characters, during the Shah's era and after the revolution, whose activities brought him to Khomeini's gallows, which helped eliminate Montazeri as Khomeini's successor.

Mehdi Hashemi, the brother of Montazeri's son-in-law (Sayyed Hadiye Hashemi) established one of the world's most powerful terrorist organizations after the revolution, with the help of Ayatollah Montazeri. So long as Hashemi's activities were concentrated on enemies of the revolution, there was no problem. Once he began to focus on members of the Khomeini regime, the seriousness of his activities was not taken lightly. When Khomeini's health began to deteriorate, Mehdi Hashemi's activities changed direction.[119] His goal–as soon as Khomeini should die–was to eliminate all of the existing leaders of the regime, including Khameneh-i (president), Musavi (prime minister) Rafsanjani (speaker of the parliament) and Ahmad Khomeini (the

Ayatollah's son). Upon Montazeri's automatic succession, Hashemi would have control over Iran and would place his own people in the government.[120] Montazeri's known choice to lead a new government, Dr. Kazem Sami, was executed by axe in his North Tehran office.

Mehdi Hashemi's intentions became known for the first time when American President Ronald Reagan attempted to initiate a dialogue with Iran, sending his former national security advisor Robert McFarlane with a cake and a Bible to see Khomeini. Hashemi wanted to kidnap McFarlane and create an international incident to prevent the establishment of relations between the U.S. and Iran. Failing at that, he revealed McFarlane's secret visit to Tehran to *Al-Sherah* a Lebanese publication. The subsequent article illuminated what was to become known as the "Irangate" scandal, and quashed the rapprochement of Iran and the U.S.[121]

Khomeini and his regime were unconcerned about the problems "Irangate" caused in the United States; what did upset them was the loss of the opportunity to secure weapons desperately needed in the war with Iraq. Hashemi was a direct threat to the existence of the Islamic Republic.

Khomeini demanded that Montazeri withdraw his support of Mehdi Hashemi and let the judicial authorities deal with the case. Since Montazeri had given Hashemi refuge in his own house, and Hashemi was the brother of his son-in-law, Montazeri refused to follow Khomeini's order. He began to criticize the brutality of Khomeini's regime in terms which made the Shah sound like a saint.[122] This behavior, along with Montazeri's sudden support of liberals (from whom he expected support after Khomeini's death in his endeavors to eliminate Rafsanjani and Khameneh-i), caused Khomeini to withdraw his nomination of Montazeri as his successor.[123]

A face-saving arrangement was made in which Montazeri

wrote a letter of resignation to Khomeini on March 27, 1989. In it Montazeri said he was not prepared for such a position and the associated burden of being a *Valiye Faghih.*[124] Khomeini accepted Montazeri's resignation in a letter dated March 28, 1979.[125] In this letter, Khomeini addressed Montazeri as *Hojat-ul-Islam* rather than as *Ayatollah*–indicating a lower rank. Montazeri lost not only his opportunity to succeed Khomeini, but his stature among the mullahs. Montazeri was kept under house arrest in his home town of Najaf-Abad for a while, and then was allowed to return to Qom, where he resumed teaching.

Montazeri's association with such dubious characters as Mehdi Hashemi did nothing for his credibility. Nor did his actions after Khomeini's death reinforce any image of his integrity. To everyone's surprise, Montazeri endorsed Khomeini's successors and encouraged the Iranians to support them. Perhaps this was out of fear of the new regime. Whatever the reason, the Iranians were probably fortunate to have been spared from Montazeri's rule.

Mehdi Hashemi was sentenced to death by the Khomeini regime after imprisonment in the notorious Evin prison. Hashemi's birthplace, Ghadarjan (near Montazeri's village of Najaf-Abad in central Iran) is noted for the cruelty of its people, a trait which Hashemi epitomized. According to Hashemi's confession prior to his execution, he had been involved in the Shah's SAVAK and during the Khomeini regime he had been involved in kidnapings, hijackings and killings.

The removal of Montazeri and the execution of Hashemi eliminated the two major obstacles to the ascension of Khameneh-i to *Valayate Faghih* and of Rafsanjani to the presidency of the Islamic Republic of Iran. The rise of President Seyyed Ali-Khameneh-i to *Valiye Faghih* surprised everyone; he was neither an Ayatollah nor a Grand Ayatollah, but only a Hojat-ul-Islam, hardly material for the spiritual leadership of the nation. Khomeini himself had been elevated from Ayatollah to Imam. There are only two ranks higher than Imam in Shi'ite Islam: the Prophet's, and God's.

Lack of rank was not a problem for Khameneh-i. According to the Islamic Republic's news media[126] Khameneh-i's succession of Khomeini came about as follows. First the Mujlis (Iranian parliament) voted against Ahmad Khomeini as his father's successor. Later, they voted 44-32 against a five-man council including Khameneh-i, Mousawi Ardabili, Meshkini, Rafsanjani, and Ahmad Khomeini. Finally, 60 of 74 members present voted for Khameneh-i as the supreme spiritual leader (Valiye Faghih) of Iran.

Khameneh-i's good fortune was due in part to Khomeini's endorsement. According to Rafsanjani[127], when Khomeini was questioned about a potential successor, Khomeini replied, "…we have plenty of spiritual leaders, but not too many of them who are also expert in the matters of internal and world affairs." Later, Khomeini suggested Khameneh-i as his successor.

Khameneh-i also owed his position to Ahmad Khomeini, who, as part of his own strategy to assume his father's mantle, allegedly forged a letter in his father's name, in agreement with Khameneh-i and Rafsanjani, which made such a step possible.[128] The letter, dated April 28, 1989, and addressed to Ayatollah Meshkini, was presented to Parliament by Rafsanjani. The letter declared that for the position of a Velayate Faghih, it was not at all necessary to be a "source of imitation" (meaning a Grand Ayatollah). Regardless of the authenticity of the letter, Rafsanjani applied his influence in Khameneh-i's favor. Elevating Khameneh-i would free up the presidency, which Khameneh-i's influence would secure for Rafsanjani. Both moves would serve to block Ahmad Khomeini, and leave the control of political power in Iran to Rafsanjani.

There is no question that Rafsanjani tricked Ahmad Khomeini. Without Ahmad Khomeini's cooperation in forging the letter, Rafsanjani could not have convinced Parliament to accept Khameneh-i as *Valiye Faghih*. On the other hand, Ahmad Khomeini's only chance to succeed his father was for the letter to be accepted as authentic; Ahmad Khomeini was himself only a Hojat-ul-Islam.

Once Khameneh-i became the *Valiye Faghih*, his supporters took no chances. His spiritual rank was immediately raised from Hojat-ul-Islam to Ayatollah. Then it was necessary to get the blessing of an old, established Ayatollah, whose endorsement would set a precedent and example for others to follow. This was given by Ayatollah Mohammed Ali Araghi, who was about ninety years old and well-known among the clerics.[129]

Khameneh-i seems to have been destined to rise in the Islamic Republic. He survived the bombing of the Parliament in June, 1981, which claimed 70 lives. (One of his arms was crippled in the blast, prompting Khomeini to call him "the living martyr.") Khameneh-i served two terms (1981-89) as president of the Islamic Republic, and when he achieved the position of *Velayate Faghih*, he was only 49 years old. There is little doubt that Khameneh-i will one day hold the rank of Grand Ayatollah, or perhaps even Imam, if his regime survives long enough.

How will Khameneh-i handle the relationship between religion and government in the Islamic Republic, now that he is in charge of the spiritual rather than the temporal side of the government. In the winter of 1987, then-President Khameneh-i stated that orders given by the government had precedence over those given by religious leaders. Khameneh-i was in trouble with Khomeini over that, until he apologized.[130] The world will have to wait and see whether similar conflicts arise between the new supreme spiritual leader and now-President Rafsanjani. Only time will tell.

It is unlikely to come to such a test, however. Khameneh-i has secured the position of commander-in-chief of the armed forces, shortly after Rafsanjani resigned the post, and thus wields both absolute spiritual as well as military power.

It has been reported[131] that Seyyed Ali Khameneh-i is a graduate of Moscow's Patrice Lumumba University. He is, and has been, a consistent supporter of Iran's cooperation with the Soviet Bloc and radical Third World states in a united front against the West. He maintains that the strategic interests of Iran and the U.S.S.R. are similar, and that this will remain so. In a speech Khameneh-i said, "The export of revolution, in the sense of exporting its values and proving the theory of the feasibility of change in Third World societies through Islamic movements of steadfastness and resistance in the face of world arrogance, must receive more attention by Muslim nations. We support all Islamic movements."

Others[132] consider Khameneh-i to be a pragmatic moderate, who has the support of the *Bazaris* (bazaar merchants) in Iran and who is interested in re-establishing an economic relationship with the West.

Whatever his real political goals, Khameneh-i has seesawed between denouncing the West and seeking to renew Iran's ties with the West, primarily for economic reasons. There is certainly no admiration lost by Khameneh-i on the West, in any event.

Khameneh-i is no Khomeini, however. Most of his success is due to his loyalty to Khomeini, dating back to 1963, when he participated in revolts against the Shah and sided with

Khomeini. The real test of Khameneh-i's loyalty came in 1973, when a theological argument divided the Shi'ite spiritual leaders. Khameneh-i supported Khomeini once again.

Seyyed Ali Khameneh-i, who came to power through favoritism and endorsement, can only be considered as a compromise, a temporary transitional figure in Iran's history. Had he relied on his talents to get ahead, he would not be Iran's supreme spiritual leader; he would not even be an Ayatollah–he failed to finish his dissertation which is a pre-requisite for receiving that title.[133]

Ali-Akbar Hashemi-Rafsanjani, former Speaker of the Iranian Parliament and now the newly-elected president of the Islamic Republic of Iran, seems to be more the hope of the Western world than that of the Iranians.[134] The West hopes that in Rafsanjani Iran will have a pragmatic and moderate leader who will gradually re-establish the political and economic relationships between Iran and the nations of the West. The Iranians, of course, hope that Rafsanjani will some-how turn their desperate economy around.

The Iranians consider Rafsanjani a wily fellow; he has managed to retain the support of both the radicals and the moderates in the Pasdaran, the army and the various branches of the government, despite criticism following the American arms deal. He was seen as clever not to have become directly involved with American officials, dealing only through intermediaries. Each step Rafsanjani took in this matter had Khomeini's approval. Rafsanjani's rapid rise from an obscure cleric with lightweight credentials is at-tributed to Khomeini's great trust in him. Khomeini believed in Rafsanjani's reliability more than any other government official, and often took Rafsanjani's advice. Khomeini's ac-ceptance of the cease-fire with Iraq was the direct result of Rafsanjani's counsel.

The first step toward consolidating Rafsanjani's power, when he decided to run for the presidency, was to initiate a plan laid with Khomeini's approval to amend the constitution. This plan would strengthen the executive, and, in particular, concentrate executive authority into the president's hands–the better to deal with political difficulties facing Iran. The office of prime minister was eliminated.

With Khomeini's establishment behind him, Rafsanjani was confident that he would be elected. On July 28, 1989, Rafsanjani was elected by 94.5 percent of the 16.4 million ballots cast. He became the third president of the Islamic Republic of Iran, succeeding Bani-Sadr and Khameneh-i.[135] Rafsanjani took the oath of office on August 17, 1989.

Shortly after Khomeini's death, but before the election, Rafsanjani visited the Soviet Union and signed several agreements which were scheduled to remain in effect until the year 2000.[136] Regardless of Rafsanjani's supposed moderation, he made it perfectly clear at a press conference with foreign journalists on June 8, 1989[137] that he would follow Khomeini's principle of "neither West, nor East" and establish relationships with any countries only on the basis of mutual respect and cooperation. Although he has tried to appear moderate on the subject of Iran's foreign policy, his conditions for resolving the differences between the United States and Iran include that the U.S. must not interfere in Iranian matters, and that Iranian assets held by the U.S. be returned. On the other hand, the U.S. demands the release of hostages held by the pro-Iranian Hezbollah in Lebanon before any deal can be made.

Breaking the Iran-U.S. deadlock may sound simple, but these three problem areas are not as simple as they seem. True, the United States froze Iranian assets in the U.S. during the hostage crisis, but these billions of dollars are either tied up in litigation or require some hard political discussions. Here are some facts:[138]

- FROZEN COMMERCE

Of the nearly $10 billion in Iranian assets transferred out of the United States in 1981, $6.1 billion was placed in escrow accounts administered by the Court of World Appeals in the Hague. The remainder was returned to Iran.

- BANK CLAIMS

Of the $6.1 billion placed into escrow accounts to pay bank claims and syndicated bank loans, $4.9 billion was paid to the banks, $494 million was returned to Iran, and $812 million remained in the accounts at the time of this writing.

- GOVERNMENT AND PRIVATE CLAIMS

Of $1 billion placed in an escrow account in N.V. Settlement Bank in the Netherlands against private and government claims, $1.27 billion was awarded in 384 claims by U.S. nationals and $118 million was awarded in four claims by Iranian nationals. Iran must replenish the account whenever the balance falls below $500 million.

- MILITARY CLAIMS

More than 2,800 contracts involving 3,800 U.S. military items remain in dispute. Iran is demanding compensation of over $11 billion; the United States says the total due is far less.

Obviously, these are complex problems that will require dialogue to resolve. Simply "releasing" Iranian assets would be pointless, especially when the two sides cannot even agree on the amount in dispute.

The negotiations for the release of Iranian assets could lead to a much bigger problem, if the solution is linked to the release of the hostages held in Lebanon. The Americans would appear to be paying ransom for the hostages, which is an unacceptable situation. On the other hand, the government of the Islamic Republic of Iran fears that giving up the hostages would be giving up their leverage in various kinds of negotiations.

The Islamic Republic's demand, renewed by Rafsanjani, of noninterference in Iran's internal matters and a relationship based on mutual respect is an even more complicated matter which requires some explanation to understand what is really meant.

If a country's economy–especially that of a Third World country–is economically and politically dependent upon the industrialized countries for the wherewithal to reconstruct and improve the economy, this dependency will bring with it some amount of exploitation. Although this is a point of concern, the real worry of the Islamic Republic of Iran and its leaders is less the threat of economic exploitation and more that the West, particularly the United States, might support opposition groups seeking to topple the mullahs' regime in Iran. In this matter, the leaders of the Islamic Republic want strict guarantees from the United States.

The same fear was manifest in the taking of 52 hostages at the American Embassy in Tehran on November 4, 1979. This occurred shortly after the success of the revolution, and exactly two weeks after the Shah arrived in New York for cancer treatments. Khomeini and his aides feared that the United States was planning some strategy similar to the one which returned the Shah to his throne after Dr. Mossadegh had ousted him. In order to prevent any such interference from the U.S., the Iranians took the U.S. Embassy personnel hostage.

Even the best of intentions will not allay this fear of the

leaders of the Islamic Republic. Even if this concern could be put aside, there is the question of how long Rafsanjani himself will survive. Since becoming president, he has survived three assassination attempts. He has also managed to remove all of his political rivals from his cabinet: Ahmad Khomeini; former Interior Minister Ali-Akbar Mohtashemi; former Information Minister Ray-Shari, who controlled SAVAMA; and former Prime Minister Mir Hossein Musavi. But the question remains as to whether he can maintain his position.

The potential for conflict between Rafsanjani and his opponents exists in matters concerning ideology and the direction Iran ought to take. Rafsanjani's faction is said to favor ending Iran's isolation and opening the economy to a freer enterprise, mainly to secure foreign aid for the reconstruction of Iran's economy. The opposing anti-Western radicals seek to tighten state control of the economy and to spread Khomeini's revolution throughout the Muslim world.

In the battle to consolidate power, Rafsanjani was the early victor. He blocked Ahmad Khomeini from succeeding his father as *Valiye Faghih*. The biggest setbacks for the opposition were Rafsanjani's removal of Musavi, Mohtashemi and Ray-Shari from their ministerial posts. The Information and Interior Ministries were two essential blocks in the foundation of Rafsanjani's power structure.

Ali-Akbar Mohtashemi was regarded as the leading purist in the government. He was replaced by Abdullah Nouri, the former government's representative in the Revolutionary Guard. As Interior Minister, Mohtashemi was publicly at odds with Rafsanjani on many issues, including Rafsanjani's stated desire to improve relations with the West. Prior to his service as Interior Minister, Mohtashemi was the Islamic Republic's ambassador to Syria and helped organize the Hezbollah Party, the pro-Iranian Muslim fundamentalist movement in Lebanon.

Western media often state that Iran has control over the Hezbollah, but these statements are vague concerning the relationship. The facts are that the Hezbollah in Lebanon was the creation of Mohtashemi, and that, as interior minister of Iran, he financially supported this terrorist organization; the Hezbollah is controlled by Mohtashemi's faction. Thus, Rafsanjani can exert little direct influence on the Hezbollah, especially since he intends to weaken or eliminate them.[139]

The seriousness of the potential threat from Mohtashemi's group, including Ray-Shari and Musavi, can be seen in the developments after Rafsanjani submitted his nominations for cabinet ministers to the Parliament. Immediately, 136 members of the Parliament signed an open letter to Rafsanjani, demanding that he restore Mohtashemi as Interior Minister. Although Rafsanjani ignored their demands, there will almost certainly be more challenges from Mohtashemi and his supporters. Many observers of Iran predict more attempts on Rafsanjani's life.[140]

True, the three deposed ministers lost their ministries and the incumbent resources, but they have not lost their influence on the hard-liners and extremists. It has been reported[141] that the assassination in Vienna of Abdul-Rahman Ghasemlou, the leader of the Kurds, was planned by Ray-Shari. Shortly after Israel's kidnaping of Sheik Abdul Karim Obeid in Lebanon, Musavi threatened revenge to the United States. Mohtashemi gave approval to the Hezbollah for the execution of the American hostage Colonel Higgins.

Further, growing evidence has been uncovered by the western media that Mohtashemi was involved in the explosion of Pan Am Flight 103, over Lockerbie, Scotland, in December, 1988, which killed 270 people. It is believed that Mohtashemi supported this action financially to avenge the downing of the Iranian airliner by the U.S. naval ship Vincennes in July of that year.

It would be naive to believe Rafsanjani is much different from his opponents. He is not a godsend for the Iranians and the West, and will not bring order to Iran and eliminate international terrorism. Rafsanjani, too, has been involved in terrorist activities and executions in the past. Since he assumed the presidency these activities continue. While Mohtashemi and his followers employ terrorism ostensibly to further the cause of spreading Shi'ite fundamentalist Islam throughout the Middle East and eventually the world, Rafsanjani's focus is on eliminating his rivals and securing his position of power. This leaves Rafsanjani little room to deviate from the goals of the hard-line Islamic fundamentalists himself; should he stray too far, he would bring himself and the Revolution down at the same time.

Rafsanjani is an opportunist, not a "pragmatist" as he is often portrayed in the American press. He wants to hold onto power and will adopt whatever policies are required to do so. He has proven himself adept at organizing coalitions with himself at the center.

If Rafsanjani is showing a moderate and pragmatic face to the world, it is simply what he knows is necessary under the circumstances facing Iran. Iran badly needs economic and technological aid. There should be no illusions concerning the fact that Rafsanjani has moved towards dictatorship in his gathering of power over the government and the Parliament.

Rafsanjani is probably the shrewdest and wiliest politician in Iran today. Although he cannot distance himself from Shi'ite fundamentalist ideology, at the same time he has a tremendous personal ambition to grasp total power. His opportunism and strong desire for power are the only hope the West has for Iran–assuming that he and his government can survive.

Ahmad Khomeini had ambitions of becoming his father's successor as *Valiye Faghih*. It has been written in various journals that Ahmad had also declared his desire to run for the presidency in both elections since 1981, but that his father forbade it.[142] Foiled by Rafsanjani in the election for *Valayate Faghih* won by Khameneh-i, Ahmad's last hope was to join forces with Mohtashemi, Musavi and Ray-Shari's opposition to Rafsanjani. At the same time, Ahmad Khomeini keeps a low profile. Perhaps he knows enough history to recognize that once a dominant leader dies, his family falls into disgrace with the new rulers. Once in power, these rulers want to cast off the shadow of the old regime.

For now, Ahmad Khomeini is one of Khameneh-i's two representatives in the Iranian National Security Council. However, this may not last long; rumors abound in Iran that a cache of ten tons of gold seized by Iranian customs officials belonged to Ahmad Khomeini.

Examination of Khomeini's legacy inevitably settles on the subject of the Iran-Iraq war. Even though the war was started by Iraq's President Saddam Hussein, the fact remains that Khomeini turned his back on the golden opportunity offered by Iraq and the other Arab nations in 1982 and underscores Khomeini's stubbornness, ignorance, short-sightedness and lack of wisdom. Wars are fought for material or territorial gain, or to depose a leader and take control of a country. In 1982, Khomeini could have achieved all of this under the peace terms that were offered. He could have demanded reparation, and perhaps even gained some territory. But, most importantly, he would have gained political advantage for Iran over the Arab nations of the Middle East, and could have parlayed that into a position of supremacy for Iran in the Middle East. How strong could Hussein have remained in such circumstances? Khomeini's best opportunity for revenge on Hussein was to accept the peace offer, not to continue the war. Khomeini would at least have gained political and economic influence over Hussein, if not the ability to remove him entirely.

The Iran-Iraq war continued for eight incredibly bloody years and stands as a monument to Khomeini for which both sides will be paying for a long time. His adversary received military and financial aid from much of the world, while his Third World country was equipped with old weapons. The Imam Khomeini cried for help from Allah. Made-in-Taiwan keys to heaven, and a seemingly limitless supply of young men to absorb Iraqi munitions, remains a tribute to the strength of Khomeini's belief in his cause. History will record that a small country destroyed the self-confidence and self-esteem of a much larger nation, due to the ignorance, stubbornness and shortsightedness of its leader: Khomeini.

After all the misery and devastation that Khomeini brought upon Iran, even in death his mission is still not finished. Khomeini left a political and religious testament for the Iranians and the rulers of the Islamic Republic to follow.

Khomeini's testament[143] was originally prepared in 1983 and delivered to the Parliament that year to be announced publicly after Khomeini's death. The document was revised and supplemented in 1987. The original document was a mere six pages; after Khomeini's death, President Khameneh-i opened the envelope in Parliament and revealed a 36-page will. Apparently Khomeini felt that the original six-page will was insufficient and too concise, unbefitting an Imam. The revised will was written in three parts, not in sequential pages, although Khomeini signed the end of each of the three parts.

Khomeini offered no new revelations in his testament. Here is a condensation of its contents:

In the first six pages, Khomeini complains that the Shah wanted to destroy Islam. He praises the Prophet Mohammed, the *Koran,* and everything to do with Shi'ite Islam as well as the victory of the Islamic Republic. Then Khomeini condemns the Americans, the Israelis, King Hussein of Jordan, King Hassan of Morocco, President Mubarek of Egypt, and President Hussein of Iraq, denouncing all as enemies of Islam.

Khomeini states that there is no higher glory than rendering the United States so helpless that they don't know what to do or where to turn. This had only been possible with the help and blessings of God. Therefore, he recommends that all the nations of the world, and Iran in particular, avoid the infidel West and the atheist East. He wants Iranians to follow the ways of the Imams, go to Friday prayers, and observe all religious ceremonies and obligations. In their prayers, the people should curse the United States, the Soviet Union and the royal family of Saudi Arabia.

The first six-page testament ends by emphasizing that this political testament is divinely inspired and applies not only to Iran, but to all Islamic nations as well as the oppressed peoples of the world, regardless of their religion.

Khomeini continues to praise God in the remaining 29 pages, saying that the revolution in Iran was a divine gift, that Islam and Islamic government are godly phenomena, and that all the prophets from Adam to Mohammed tried hard and sacrificed to achieve it. He adds that since this divine gift was given to Iran, and since he is already taking the last breaths of his life, he wants to make recommendations concerning protecting this gift for future generations.

   1)The secret of the victory of the Islamic Revolution was the unity of words. This must be kept further. Division among you must be avoided.

2)Islam is not against science and technology. Islam is not separate from politics. Islam is only against the infidels and prostitutes.

3)There are those who say that the Islamic Republic did not do anything for the people; it is not true. If the opponents would be fair enough, they would agree that we have done plenty for the people who fought for the faith during the time of the Prophet Mohammed. Therefore, this government must be protected and it should become an example for other Islamic countries.

4)This was a result of the colonization policy to prevent the clerics from participating in the education system of the country and therefore a division between the clerics and the university professors was created. The professors and the clerics must get closer and cooperate with each other.

5)Our nation has been impressed for years by the way of life in the East as well as in the West, whereby our Aryan race and the Arabs should have nothing less than the Europeans, Americans and Russians. If they realize that and don't pay any more attention to others, they can achieve everything.

6)My command to the honorable nation of Iran is to participate in all elections, whether for electing the president or the members of the parliament. Keeping away from elections is counted among the biggest sins. Also my command to the ruling group is to devote themselves to serving Islam and the oppressed people. They should not think that being part of the ruling elite is a present for them from God. They have a very heavy duty and not fulfilling it will carry shame for them for eternity.

7)Justice is a necessity which involves people's lives, property and chastity. My command is not to give any position in matters of justice to unqualified people.

8)All the centers for spiritual teachings, especially Qom, must protect themselves from getting involved in any activity which could cause harm to Islam.

9) My command to the executive legislators is to serve the nation, because the ministers can only remain in their positions if they serve the people. The Foreign Ministry must try to rescue the nation from becoming dependent on other nations. The Guidance Ministry must try to introduce the real face of the Islamic Republic.

10) One of the most important topics is the matters concerning the kindergartens up to the universities. My command to the youth is to protect themselves from any kind of deviations.

11) My command to the army and other forces is not to be deceived by the enemies of the country and not to try to get rid of the Islamic Republic, and they should avoid getting themselves involved in politics.

12) My command to the Parliament, the President and other government institutions is to prevent the press and media from deviating from Islam's interests and the interests of the country.

13) My advice to all the groups and persons opposing the Islamic Republic is that you cannot bring down any government with your inhuman methods. I advise you that you should stop your nonsensical activities wherever you are and you should not be deceived by others. If you have not done any criminal act, then you should return to your country and the lap of Islam and you should ask for pardon. Because God is merciful, the Islamic Republic and the people will forgive you. Don't waste your life any further. My duty was to guide you. It is my hope that you will listen to me. My command to you and all the parties and groups, whether you are leftists or those Kurds and Beluchis, who in the name of autonomy are fighting against the Islamic Republic, is to be sure that Islam is much better than the Western criminals and the Eastern dictators, and it meets the needs of the people much better. My advice to the writers, intellectuals, speakers and critics is that if you believe in God, then get in touch with your God, and if you don't believe in God, then try to get to your conscience and judge with all fairness whether it is fair to

fight someone who has rescued the country from the hands of foreigners, has closed all the centers of prostitution, bars and pleasures. We have served this country. The big difference between the former regime and the Islamic Republic is that the Shah's regime used to care about the rich people and the cities; we, on the contrary, concentrated more on the villages and the oppressed people.

14) A matter of which I would like to remind you is that Islam neither agrees with the cruel and exploiting system of capitalism, nor with communism, which is against private ownership. My command to the governing body of the country is to keep away from these two poles, and to concentrate more on the just Islamic teachings and develop the farms, villages and factories.

15) My advice to the opposing group of clerics is to stop fighting the Islamic Republic, because first God would not let them reach their goal. On the other hand, if the Islamic Republic is defeated, nobody will follow you and you will regret your acts.

16) My command to the Muslims and the oppressed people is not to wait until the liberation bears fruit. You must get up and get your share. You should not be scared by the propaganda of the superpowers, because the final victory is yours.

17) Once more, I command the nation of Iran that in the name of almighty God, you should move towards getting back to your self-recognition, and reach autarchy and liberation with all their benefits. You should move forward and without doubt that God is with you, and I can leave you with a sure heart and a happy soul and go to my eternal peace. I need your prayers. I beg from the merciful God to forgive me for my shortcomings in serving you. For the nation, I hope that they move forward with a strong will. The nation should know with the death of a servant there should not be a split in the iron will of this nation, and there are more servants who are even better, and God is the protector of this nation and the oppressed people of the world.

At the end of the will, the final page touched on matters more personal to Khomeini:

1) I want to make clear that anything said or written in my name is only valid if it carries my signature, or is documented through my voice or my person speaking it.

2) Those people who claim to have been writing my communiques lack any truth, because all the communiques were written by myself.

3) Some people have claimed that my trip to Paris took place upon their recommendation; this is baseless. After I was expelled from Kuwait, I chose Paris in consultation with Ahmad. The reason is that no Islamic country would have allowed me entry into their country, since they were under the Shah's influence, but France was not.

During the revolution I praised many people who I later found to have deceived me. These praises were made because they showed themselves faithful to the cause of the Islamic Republic, and later it turned out to be that they were taking advantage of it.

Khomeini's will leaves two distinct impressions upon the reader. The first is that Khomeini strongly believed that the Islamic Republic was the salvation of the Iranians. And Khomeini was fearful that the regime which he had set up could not last, and could be brought down at any time.

Surprisingly, there is evidence that the good Ayatollah's faith was waning as he neared the end of his life. A few weeks before his death he wrote a poem about waking up disenchanted with schools and mosques, and wanting to drink wine and go to bars.[144]

Khomeini's revolution in a country which was well on the way to becoming a modern and prosperous nation, brought unimaginable devastation. The Iranians can only hope that the mullahs' regime is as vulnerable as Khomeini feared.

10

THE ISLAMIC ECONOMIC CONCEPT:

A Solution?

When Khomeini first set out to win the support of the Iranian people, they were told that the Shah and his regimes had exploited them and their country. With the huge profits from oil, Khomeini went on, no one should suffer economically any longer. He promised that when the Islamic Republic replaced the monarchy, every Iranian would receive 500 rials per day. This was a very enticing promise for the average Iranian, who is ignorant about economics. For a household of five, this would mean 75,000 rials per month, about twice what the head of the household would make by hard work; it didn't take much to get these people on Khomeini's side.

Khomeini came to power, and the Islamic Republic replaced the Shah's regime, but as yet no Iranian has received a check for 500 rials a day. Instead, almost overnight Khomeini's regime brought economic misery as had not been experienced in over twenty years.

During the Shah's era, the economy of Iran, despite its shortcomings, boomed like never before. There was tremendous economic growth, with full employment to the extent that Iran hired hundreds of thousands of "guest workers" from Pakistan, India, Afghanistan, and Korea along with tens of thousands of experts from the industrialized nations of the world. Inflation was high compared to the standards of the West, but nothing like it would be under Khomeini. Income from oil exports reached new highs, with $24.3 billion in 1977 and $22.5 billion in 1978. In short, despite unjust distribution, Iranians enjoyed full employment, and earned enough money to buy the goods and services they needed, and could afford luxury items as never before or since.

Khomeini's impact on the economy was exactly the opposite. Here are a few of the Khomeini regime's accomplishments:

- Full employment rapidly declined, and today there are at least four to six million people unemployed in Iran[145]; some estimates run as high as 20 million.[146]
- Inflation rages in the 400% annual range.[147]
- Income from oil exports dropped to $11.1 billion in 1987.
- Economic losses from the war with Iraq are estimated to be $627 billion for Iran (and $561 billion for Iraq).[148] Even if the actual figures are only one-tenth of those estimates, the costs to the two shattered nations are staggering.

The effect on the daily life of the people has been devastating. Today there are shortages of everything. Bread, meat, eggs, cheese, sugar and cooking oil are worst of all; people can only get sufficient quantities of these staples on the black market, at prices which few can afford. Rather than liberation and plenty, the Khomeini regime brought the people oppression and deprivation on a grand scale.

In an attempt to keep the situation under control, and to keep the black market from getting out of hand, the Islamic government distributed ration coupons for primary commodities such as food items. Apparently this was considered the fairest method of distributing wealth to the oppressed and deprived people, whose ranks swelled enormously under the Islamic government.

The economic record of the Khomeini regime is on a par with its achievements in human rights.

As long as the war with Iraq continued, Khomeini could demand sacrifices by the people. Now that the war is over, and Khomeini is dead, the new leadership in Iran recognizes the desperate need for economic reforms to save the devastated country from total ruin. Will the Rafsanjani government be able to turn the economy around? Will Islam's unique economic concept prove helpful, or doom the government's effort to failure?

The near-term potential for economic development is bleak. Even if the West immediately began cooperating with Iran to rebuild the economy–which is highly unlikely at this time–and regardless of several economic agreements with the Soviets, the Islamic government faces two probably insurmountable obstacles: shortage of trained personnel and critical experts, and lack of funds to finance the smallest steps to recovery.

The Shah's regime had made a long-term investment in the education and training of Iranians to fill management and leadership positions in industry and government. At the time of the revolution, Iran had well over a million well-trained and highly educated people in charge of its factories, refineries, oil fields and infrastructure. When the revolution came, over three million Iranians fled their country, most of whom were the skilled people any modern economy needs to survive. Educated people were, after all, a threat to the Islamic government.

A great number of formerly productive workers found themselves in military service in the war with Iraq, in which a million or more were killed and half a million crippled. Hundreds of thousands of people were imprisoned in revolutionary fervor. Most of the remaining workforce, mostly unskilled laborers, has been unemployed for so long that simply returning to work will require a major adjustment. Even if the Islamic government could manage to organize the limited human resources it has, capital to do anything with them is sorely lacking, and likely to remain so; the Islamic Republic has few friends these days.

A primary requirement of any economy is that the citizens believe in it and cooperate with it. The past actions of the Islamic government have done little to engender trust on the part of its people, and there is little reason to believe that the administrations of Rafsanjani or his foreseeable successors will be any more successful in this fundamental respect.

The Islamic Republic is doomed to economic failure. Either it must adhere to the Islamic economic concept, which is derived from the Koran and inappropriate in the real world, or it must deviate from its Islamic economic path, which would bring about the end of the Islamic Republic.

To understand Iran's dilemma, one need only consider Islam's concept of economics. This concept is based on the verses of the *Koran* wherein Mohammed spoke of matters relating to individual business and state economics. Recently, Islamic economists who have studied the modern economic theories of Smith, Keynes, Marx and others, have tried to reconcile Islamic economics with modern theory. Recent Islamic economic writings often use *Hediths* (narrations) to explain how certain circumstances should be handled; the *Hediths* explain how similar situations were handled by the Prophet Mohammed and the Caliphs.

Muslim economists have been writing on the subject practically since Islam began. The most famous–and prolific–of them was Ibn Khaldun (1332-1406).[140] Ibn Khaldun addressed many aspects of economics, including the concepts of value, division of labor, the price system, the law of supply and demand, consumption and production, money, capital formation, population growth, public finance, trade cycles, for example. He discussed the various stages of economic progress in societies. Even Schumpeter made reference to Khaldun in his compendium.[150]

Other famous Muslim economic thinkers were Abu Yusuf (731-798), Nasiruddin Tusi (1201-1274), Ibn Taimiya (1262-1328), and Shah Waliullah (1702-1763).[151] These men based their thoughts on the verses of the *Koran*.

Although there is no shortage of Islamic economic teachings, which have been practiced to a limited extent various countries, there is no empirical data to show what would happen if a nation were to adopt and practice the Islamic economic concept as preached in the *Koran*. Still, the Islamic economists assure us that their system is much more just than either capitalism or communism[152].

A totally Islamic state is described by Dr. Abdul Quader Shaikh, a Professor of Business Economics, as follows:

> [An] Islamic state is defined as one where the principles of Islam, as dictated by the *Koran*, Shariat[153] and Sunnah[154] are followed. It can further be visualized in this context as a society or state which is regarded as a state, responsible for

providing living wage relief to every inhabitant, i.e. whether Muslim or non-Muslim; in which there are neither slums nor millionaires, neither the exploited proletariat nor the exploiting bourgeoisie. It is a society where there is no accumulation of wealth in the hands of few, absence of hoarding or profiteering, since both are taboo in Islam. Profiteering is considered with dismay even in corporate capitalistic systems, though profit-making is accepted [in these systems] as a normal course of business.[155]

The realization of an ideal Islamic state, according to the Muslims, would be achieved through the social order of Islam, which embraces spiritual, moral, political and economic measures.[156] For the sake of brevity, the discussion here will concentrate on the economic aspects of Islamic doctrine.

According to Islamic economic literature[157], an Islamic economy is built upon two fundamental principles: *Zakat*, and the prohibition of interest. Here is a quotation from Islamic economic literature, describing *Zakat* and the economic functions and goals it should fulfill; the spiritual and moral aspects of *Zakat* are not discussed here.

> *Zakat* is not charity. It is the right of the have-nots in the wealth of the haves. Also, it is something entirely different from taxes levied for meeting the cost of government. It is a measure designed to transfer part of the wealth from the haves to the have-nots in society. *Zakat* covers almost every kind of wealth: cash and bullion, stocks and inventories, agricultural produce, cattle, mineral wealth, etc. Exemption limits and rates have been fixed once forever. As a permanent provision in Islamic law, this redistributive measure is necessitated by the unequal opportunities. This measure is supplemented by the Islamic law of inheritance in the limited sphere of the family and kinfolk.[158]

The rationale for the Islamic prohibition of interest is as follows:

The main reason why Islam abolishes interest is that it is oppression involving exploitation. In the case of consumption loans it violates the basic function for which God has created wealth, which envisages that the needy be supported by those who have surplus wealth. In the case of productive loans, guaranteed return to capital is unjust in view of the uncertainty surrounding entrepreneurial profits.

The second reason why interest has been abolished is that it transfers wealth from the poor to the rich, increasing the inequality in the distribution of wealth. This is against the social interest and contrary to the will of God, who would like an equitable distribution of income and wealth. Islam stands for cooperation and brotherhood. Interest negates this attitude and symbolizes an entirely different way of life.

A third reason why interest is abolished is that it creates an idle class of people who receive their income from accumulated wealth. The society is deprived of the labor and enterprise of these people. Such a way of life is also harmful to their personalities.[159]

The moral and distributive aspects of *Zakat* and the prohibition of interest are praiseworthy; in the case of the latter, it also could make a lot of economic sense. However, these two concepts are not without shortcomings which render them either ineffective or counterproductive:

1) *Zakat* is, from an economic standpoint, a distributive rather than a productive measure; a kind of taxation of income. Results of decades of capitalism–as well as communism or socialism–clearly show that any taxation of income discourages investment and has a dampening effect on economic growth. This is the primary reason that "supply-siders" recommend tax reduction in order to stimulate an economy. When the productive side of an economy is neglected or exploited, the distributive side will fail sooner rather than later.

2)The abolition of interest makes tremendous economic sense, regardless of the moral and distributive aspects of prohibiting interest. The phenomenon of interest causes an imbalance between production and consumption. This happens in two ways. First, interest on consumption loans transfers part of the purchasing power of a group of people with a high propensity to consume, to a group with low propensity to consume, for example, a group with more wealth than they can consume. This group reinvests its interest income which means that the decline in consumption is accompanied by an increase in production. Second, interest on productive loans raises costs of production, and hence the prices on consumer goods. Once again, the amount of this "tax" in the form of higher prices transfers wealth into the hands of a class with a lower propensity to consume. This imbalance manifests itself as stagnation, depression, monopoly and ultimately imperialism.[160]

The underlying principle in the abolition of interest is that Capital should not be more highly valued than Labor; Capital and Labor should be equal factors in production. This is also a fundamental difference between capitalism and communism. Capitalism values Capital more than Labor, and communism takes the opposite view. Islam sees Capital and Labor as equal in importance.

However, since the early 1970s, every economy in the world, regardless of ideology, has been hit with shortages or increasing costs of energy and raw materials, as well as compounding environmental problems. Capital and Labor are not the only factors which must be considered for production any more. Under today's circumstances, production is a function of a combination of capital, labor, energy, raw materials, protection of the environment, and technology (costs of research and development, or the transfer of know-how).

Abolition of interest and the establishment of interest-free economies in all of the Islamic countries–most of which belong to the Third World–would be ineffective, even counterproductive. The most that these Third World countries could accomplish would be to make their few wealthy people join the ranks of the masses–as have-nots. The countries would become even poorer and even less capital would be available for productive purposes.

It is important to remember that before anything can be distributed and justice and brotherhood attended to, it is imperative that something be produced. Diffusing Capital does not promote production.

3) Another way in which Islamic principles thwart any possibility of economic success in Islamic countries is the fact that half their populations–women–are excluded from participation in the productive side of the economy. Women are not allowed to produce, yet they necessarily consume. This guarantees shortages and prices which are too high as too little wealth chases even less available goods and services.

4) This leads to another reason for the backwardness of Islamic countries: little effort is made to raise the level of public education. In a country ruled by the Islamic clergy such as Iran, education would be dangerous to the health of the government. Of course, high illiteracy rates prevail among all Islamic populations, but even less attention is given to the "inferior" half of the population, the women. This, of course, affects future generations, because the mother is the first source of learning for every child. Ignorance will condemn these societies to remain economically backward.

5) The Islamic notion of promoting brotherhood and justice by distributing the wealth of the haves to the have-nots cannot create full employment, control inflation or initiate sound economic growth.

The flaws of the Islamic economic concept become more evident when considered in the context of heavily-populated countries. Even oil income cannot support the economy of such a nation, as has been demonstrated by the Islamic Republic of Iran.

At best, the Islamic economic concept might be said to work in a few of the oil-rich, underpopulated Arab countries such as Kuwait, Saudi Arabia, Libya and the United Arab Emirates. Even so, if these oil rich countries start sharing their wealth with their have-not Islamic brethren, they will find that brotherhood and justice according to the *Koran* will put them in the same straits.

One cannot overlook the potential for popularity of the Islamic economic concepts, and its potential for political power. No doubt such ideas as *Zakat* and prohibition of interest in the spirit of brotherhood and justice would be enticing to poor and deprived people anywhere.

The Chinese philosopher Kuant-Tzu once said: "Should you give a man a fish, he would feed himself for one time. But should you teach him to fish, he will feed himself for the rest of his life." To this could be added, if you take the fish away from a successful fisherman and give it to those who do not produce for themselves, the fisherman will not keep fishing for long, and then everyone will starve.

11

THE AVENUE TO A STABLE AND PEACEFUL IRAN

*I*t would not be a simple task to come up with a solution to all of Iran's miseries which would be acceptable to all concerned, especially given the diversity and radicalism of the various political groups involved, each struggling to realize its own goals. However, suggestions which are based on logic and fact should not be discouraged, especially those which consider the general best interests of Iran and the Iranian people, rather than simply the interests of a single political group.

The main focus in this chapter will be to offer such a solution. In so doing, however, an attempt will be made to correct a distorted view of the Baha'is which has been propagated by the mullahs for decades. Through false accusations, brutal executions and other barbarous handling of the Baha'is, the mullahs have deprived Iran of the contributions of thousands of well-educated, peace-loving people who could have done so much for their country. The only beneficiaries of this brutality were the Shi'ite mullahs in Iran.

Before proposing the remedies for Iran's miseries, a review

must be made of the factors which contributed to these miseries over the centuries:

- The conquest of Iran by the Arab Muslims, beginning in 642 A.D. and ending in victory in 651, imposed Islam and its teachings upon the Iranians. This event virtually halted progress and led to the domination of Iran by the Shi'ites and their mullahs. As a result, Iran lost its identity and became the victim of a backward ideology, from which it has suffered ever since.

- The Shi'ite mullahs of Iran became the long arm of the colonial powers and other foreigners with exploitive intentions, most notably Great Britain.

- The Freemasons established a major stronghold in Iran, influencing Iranian politicians and using them to carry out their policies.

- The governments of foreign countries, such as Great Britain, the United States, the Soviet Union, France and Germany realized their goals in Iran through their hirelings in various religious and political groups of all ideologies.

- Multinational corporations, primarily the major oil companies, imposed their economic will upon Iran with their exploitive policies over the years.

The list is endless. However, these examples are sufficient to illustrate the development of disunity among the Iranians over the years. Each faction's members became the pawns of the different foreign interests whom they served.

The effects of these foreign interests penetrated Iran's population so deeply that Iran, although a sovereign nation, was in fact disintegrated, its people splintered.

By identifying the problems and their sources, a three-part approach to the reunification of Iran's people is made clear. Foremost is the replacement of Islam, an Arabic religion, with an Iranian religion, one that originated in Iran and is compatible with Iranian ideology. This measure would free the Iranians from the fetters of slavery to foreign ideologies, and provide an effective means for Iran to impose its peace-loving and progressive ideology on others. This single measure would benefit the people of Iran in various ways.

The second part of the solution is the total removal of the mullahs from Iran, a complete dismantling of their institution, and the creation of a system which would guarantee that they could never be resurrected. This would help to remove foreign influence from Iran, and to open the door for the government to be run by qualified people acting in the interests of the country.

The final part of the solution is the elimination and prevention of all external influences imposed upon Iran, either directly by organizations such as the Freemasons, or indirectly through the establishment by foreign powers of political or ideological groups in Iran.

Should these objectives be realized, Iran would be well on its way to the most glorious era of its history. The question is, how could these measures be implemented?

To date, most of the Iranian people have overlooked the greatest event in the history of Iran: the birth of the Baha'i religion in that country. The Baha'i faith is an Iranian ideology, religion and movement, which has the potential to solve the problems which have plagued Iran. The teachings of the Baha'is are progressive, and designed to unify mankind and to attain the "Most Great Peace" in the world of humanity.

Iran occupies special status in the Baha'i faith, because Iran is the country of origin of this emerging global religion. Since its recent origin (1844) the Baha'i faith has spread to practically every nation on earth, with millions of practitioners. The Baha'is consider Iran to be a holy place and the center of the world. In a Baha'i Iran, foreign influences would have no room to maneuver. Moreover, Iran would become the center for pilgrimages by Baha'is from the world over. Iran would once again be a place that people would want to visit.

An interesting, and appropriate, feature of the Baha'i faith is that it has no clergy of any kind; the Baha'i tenets strictly forbid it.

By considering these points, one can realize that proliferation of the Baha'i faith would solve much of Iran's woes. This would require the majority of the Iranian people to become Baha'is, which is not impossible. The Baha'is are already the largest minority in Iran, outnumbering the Sunni Moslems, and most Baha'is converted to the faith themselves. Among the many Muslim converts to Baha'ism are numerous former mullahs, many of whom persecuted the Baha'is until they became aware of the Baha'i teachings. The number of Baha'is in Iran increases daily. Baha'ism has spread rapidly because it makes sense to so many people of other faiths; Baha'ism actually fulfills a Shi'ite prophecy.

Should Iran experience the spread of the Baha'i faith, Iran could be a springboard to the world for the progressive and peace-loving teachings of Baha'ism. Although the teachings and tenets of the Baha'i faith would lead to a united Iran promoting unity and love for mankind, the Baha'i doctrine has been completely misrepresented to the people by the mullahs of Islam since its inception.[161] The mullahs have managed to generate such animosity toward the Baha'is on the part of the fanatical Muslim population that in the brief history of Baha'ism, tens of thousands of its followers have been persecuted, killed, imprisoned or executed. The Baha'is who escaped direct prosecution were deprived of all the basic rights to which an Iranian citizen is entitled.[162]

In their efforts to turn the Iranian populace against the native Baha'is, the mullahs have been successful to the point that the Iranian people no longer know their saviors from their enemies. It is also surprising to note that, despite the fact that the mullahs are corrupt, they have brought misfortune to Iran under cover of Islam. Many of the mullahs are agents of foreign powers. If one were to point to the troubles in Iran and allude to the reality there, the Iranian Muslims would become defensive and protective toward their mullahs. Such is the power of the mullahs' propaganda. There is no doubt that the mullahs, in pursuit of their own political and economic goals, have thoroughly brainwashed the Muslim people of Iran.

Numerous baseless accusations have been made against the Baha'is; some will be presented here. There have been many publications involving anti-Baha'i propaganda. One considers the Baha'i faith to be an extension of Czarist Russia's exploitive policy in Iran.[163] That book presents an absurd and illogical thesis, supported by the paramount truth that the first Baha'i temple is located in Ashg-Abad, in the Soviet Union.

At the same time the mullahs proclaim that the Baha'is are the long arm of Zionism, since the Baha'i World Center is located in Haifa, Israel. Wherever the Baha'is build a new center, they are accused by the mullahs of conspiracy with that country.

According to another publication[164], Baha'ism is an offshoot of the Shi'ite sect. This work presents misconceptions and misinterpretations, with false quotations and photographs whose captions misidentify the people in the pictures. There are numerous similar publications[165] whose sole purpose is to degrade and discredit the Baha'i faith by misrepresenting it to the Iranian people.

Before considering the details of what the Baha'i faith represents, it will help to begin with the Baha'i perspective in the following questions: Given that many of Iran's problems stem from religion, or in the name of religion, is religion necessary? Should religion and politics be kept separate from each other? If Baha'ism were to become the dominant religion in Iran, what assurance is there that things would be any different than they are today?

There are three aspects to every religion: the philosophical, the psychological, and the sociological, which are of concern. The philosophical and psychological aspects of a religion are designed to strengthen the faith of the followers, to create hope for them on the earthly plane as well as in the world beyond, and at the same time to acquire a hold on people. As a consequence, the stronger the faith of the followers, the stronger the control of the religious leaders. These two aspects of religions are manifested in having the faithful chant prayers, read religious narratives, participate in ceremonies, and follow religious teachings and customs. Karl Marx was not wrong when he stated that religion is the opiate of the masses.

The sociological aspect of religion includes the religious rules and customs which are made to govern the lives and relationships of the people in society. These are marriage and divorce, education, basic conduct, administration of punishments, for example. Since the state and religion both try to regulate society, this is where competition between religion and the state begins. Should the social laws enacted by the state differ from the religious tenets of a predominantly religious society, the hostility between religion and the state will be proportional to their differences. On the other hand, in a country strongly influenced by religion, much more religious regulation is imposed upon the people. Either way, religion interferes with politics, as a result of which it becomes impossible to separate religion from politics. In fact, there is not a non-communist country in the world which is not governed by the direct or indirect influence of religion; this is true in many communist countries as well.

The conclusion to be found in this brief analysis is that religion is necessary for people in general, and for Iranians in particular. Since the interference and influence of religion in politics is unavoidable, it would be wise to adhere to a religion whose teachings and rules are compatible with a modern and progressive society. The influence of such a religion on society could only be positive. Therefore a separation of religion and the state, which is practically impossible anyway,[166] would not be required.

In comparison to all of the other religions in the world, Baha'ism seems a Godsend to the Iranians, which could enable them to solve their internal problems, and could lead Iran to a glorious future.

An understanding of the special nature of Baha'ism begins with the basics of religion in general. Religion has spiritual aspects, and social ones. Although they differ in detail, the spiritual aspects of most religions are essentially alike. All involve communication with some form of deity through prayer, and the promise of some sort of life after death. This side of religion is less relevant from the political standpoint, because its effects of the adherents of any religion is more or less the same.

The social side of religion is what makes the big difference. The teachings of a religion which is modern, progressive, peace-loving and a source of unity could be a constructive influence on a society. A religion with radical and repressive teachings can lead a country to destruction.

A big mistake of would-be progressive leaders, such as the Shah of Iran, is their belief that they can create a modern society by importing technology, building factories, and acquiring automobiles, computers, televisions, airplanes, tall buildings and modern weapons. Possession of the trappings of modern life does not make a country fully developed. A progressive, modern society is attained only when its people enjoy human rights and dignity, peace and security, in addition to economic prosperity. By their ideology and their actions, the mullahs have demonstrated that the Islamic Republic of Iran cannot become a progressive society as long as it is Islamic.

The Baha'i faith was founded in Iran by Mirza Husayn Ali (1817 - 1892), also known as Baha-u-llah (the Glory of God). The word Baha'i denotes a follower of Baha-u-llah. The origin of the Baha'i faith is the Babi religion, which began in Shiraz, Iran, in 1844, founded by Seyyed Ali Mohammed Shirazi, also known as the Bab (the Gate), who emphasized the forthcoming appearance of a new prophet or messenger from God.

The Bab himself, according to the Baha'is, was the fulfillment of the prophecies of Judaism, the long-awaited messiah. And the Bab was also the Twelfth Imam, the Hidden Imam of the Shi'ites.

Although the teachings of the Bab, whose principle theme was the promise of the comings of "Him Who God Shall Make Manifest," spread throughout Iran and gained a large number of followers, they garnered strong opposition from the Shi'ite mullahs and the government. As a result, the Bab was arrested, and after several years of imprisonment, was condemned to death in 1850. After the execution of the Bab, large scale persecutions of his followers began, in which more than 20,000 people were killed. Many of these executed Babis were former Shi'ite mullahs, who had embraced the Babi religion after the Bab's proclamation.

The execution of the Bab provoked two Babis to make an attempt on the life of the Iranian monarch, Nasirid-din Shah, to avenge their fallen leader. Even though this attempt was unsuccessful, Baha-u-llah, who was at this time an avid follower of the Bab but had no knowledge of the assassination attempt, was arrested and thrown into the Black Pit, a notorious jail in Tehran. It was during this imprisonment that Baha-u-llah was made aware of his mission as God's messenger on earth. After a year of incarceration, he was released in January, 1853, and subsequently exiled to Baghdad, Iraq.

The public proclamation of Baha-u-llah took place in April, 1863, during his stay in Baghdad. He declared to a small group of Babis that he was the Messenger of God whose advent had been prophesied by the Bab. This public proclamation revived the Babi community, but it also alarmed the Persian government, which urged the Ottoman government to move Baha-u-llah and his growing number of followers to Constantinople, farther from the Iranian border. After a short sojourn in Constantinople, Baha-u-llah was transferred to Adrianople. In this city, Baha-u-Llah made the second public proclamation of his mission. This time it was in the form of tablets addressed to the rulers of Iran, Turkey, Russia, Prussia, Austria, and Great Britain, as well as to the Pope and to the Christian and Muslim clergies collectively.

Baha-u-llah's last exile was to Akka, in Palestine, where he died in 1892. By this time, not only had the majority of Babis acknowledged Baha-u-llah's claims and become Baha'is, but the Baha'i faith had spread beyond Iran and the Ottoman

Empire to the Caucasus, Turkistan, India, Burma, Egypt and the Sudan.

Baha-u-llah appointed his eldest son, Abdul-Baha ("Servant of the Glory", 1844-1921) as the leader of the Baha'i community, and authorized him to interpret his teachings. Abdul-Baha administered the affairs of the Faith from Palestine, and also actively engaged in spreading his father's message in his travels through Africa, Europe and the United States from 1910 to 1913. During his ministry, Baha'i groups were established in North Africa, the Far East, Australia, and the United States. The Baha'i faith was spread to practically every country in the world.

Abdul-Baha, in turn, appointed as his successor his eldest grandson, Shoghi Effendi Rabbani (1897-1957), who became the Guardian of the Baha'i Faith and the authorized interpreter of the teachings of his great-grandfather, Baha-u-llah. During Rabbani's tenure, the message of the Baha'i Faith reached the farthest communities in the world. (The oft-cited phrase "New World Order" was coined for the first time on November 28, 1931, by Shoghi Effendi Rabbani as the title for a book he had written.)

After the passing of Rabbani, the Universal House of Justice took charge of the Baha'i communities of the world, and functions as the supreme administrative, legislative and judicial body of the Baha'i Commonwealth. The seat of the Universal House of Justice is in Haifa, Israel, near the shrines of the Bab and Abdul-Baha, and near the shrine of Baha-u-llah at Bahji, near Akka.[167]

An overview of the teachings of the Baha'i Faith must begin with a summary of the teachings of the Bab.[168] The main points of the Bab's teachings were:

- Preparing human beings for the acceptance of a great change in the civilization of mankind.

- The declaration of the end of the Islamic era and that of prophethood.
- Glad tidings to the world for the creation of a new order.
- Rejection of the clergy and the clerical system, as well as of the concept of theocracy.
- Rejection of religious jurisprudence over the people.
- The liberation of women from unjust treatment.
- The necessity of establishing a system of government of the people by the people themselves.
- Explanation of the concepts of heaven and hell, relating them to this world; considering the world beyond as a spiritual world.
- Rejection of all religious titles and positions, such as Faqih, Pope, Ayatollah, which he considered chains fastened to the legs of the people.

It is obvious why these teachings and principles outraged the Shi'ite mullahs and inspired their desire to stifle and destroy the movement in its embryonic stage.

The teachings and principles of the Baha'i Faith are further developed and refined. The twelve major principles of the Baha'i Faith are:

1) The oneness of the world of humanity.
2) The foundation of all religion is one.
3) Religion must be the cause of unity.
4) Religion must be in accord with science and reason.
5) Independent investigation of truth.
6) Equality between men and women.
7) The abandonment of all prejudices among mankind.
8) Universal peace.

9) Universal education.

10) A universal auxiliary language.

11) Solution of economic problems.

12) An international tribunal.

The Baha'i Faith is based upon three main areas of belief: the existence of God; that the prophets are sent by God to guide mankind and that the prophets forward the message of God to mankind; and that after death the soul lives on. These three beliefs are common to most religions.

The Baha'is believe that God sends messengers from time to time to guide the people. Since the spiritual and social teachings of each prophet are suitable for his own time and the capabilities of the people of that time to understand them, as time passes and mankind progresses, new teachings are necessary.

The beauty of the Baha'i religion is to be seen in its goal to unify humanity, regardless of origins, nationality, sex, race or color. This unique characteristic of a religion manifests itself for the first time in the Baha'i Faith. No other religion accepts the right of other religions to exist; in fact, most religions cannot achieve unity even among their own followers. While it is one thing for the Muslims to consider the Christians and Jews as infidels, and for the Jews to deny that Jesus Christ was the messiah, it is something else for the followers of the same faith to quarrel with one another. The split of the Sunni and Shi'ite Muslims is a prime example of disunity of the Muslims. The same phenomenon can be seen in the various sects of Jews and Christians, and of the other religions which preceded Baha'ism.

The foundation of the Baha'i Faith was laid down in such a manner that there will never be disunity among its followers, because the Baha'is believe wholeheartedly that religion must be a source of unity and love among mankind. The abandonment of all prejudices and the equality of men and women, as well as the solution of economic problems, are but a few of the important steps designed to achieve dignity and freedom for all mankind. According to Baha-u-llah himself: "Should religion be the cause of anything else but unity amongst mankind, it is better that you have no religion."

From the Baha'i point of view, the solution of the world's economic problems can be accomplished by the elimination of the extremes of wealth and poverty. However, the Baha'is do not advocate communism or socialism. Rather, they favor a system of "General Capitalism" as a system which is fair to everyone. More importantly, it should lead to general prosperity. To achieve this goal, the Baha'is primarily emphasize education for each and every individual, putting them in a position to become economically self-sustaining and independent from charities, Zakats and things of that nature.

In considering education as the key to individual as well as collective economic strength, the Baha'is pay special attention to the education of women. In their view, if parents can afford education for only one of two children, the girl should be taught, because she will be a mother and source of education for the future generation. The emphasis on education has stood the Baha'is well. The Baha'is during the Shah's era were generally well off, and this triggered the animosity of the mullahs.

With the creation of an international tribunal and the adoption of a universal auxiliary language, the Baha'is do not intend to abandon national governments. Rather, they seek to establish a forum empowered to resolve disputes and prevent conflicts and wars between nations. These and other principles of Baha'ism are intended to attain universal peace.

The Baha'is believe that one day, in the not too distant future, Iran will become the most prosperous nation on earth, and the envy of all the world. The main factor which will make Iran the envy of the world will be its primacy in the hearts of the Baha'is. Someday vast numbers of Baha'is will pour into Iran, making the pilgrimages denied to them during the reign of the mullahs. Not only will they bring lots of foreign currency, but they will help Iran regain its lost credibility among the family of nations, and will bring back the respect of the world for Iran and the Iranians.

Although the Baha'i Faith seems well suited to lead Iran into a glorious era, the question remains whether the involvement of this religion in politics would turn out any differently than that of others. To appreciate the great difference in the Baha'is in this respect, one must first consider the nature of governments and how they are put together.

Two situations are commonly observed in the formation of governments: either the will of a minority is imposed upon the majority, or the majority dominates the minority. In the first case, the government is usually either a monarchy, a dictatorship, or some form of totalitarian regime. Although in the second case the government could be a democracy, in a democracy the majority can still override the minority by even a slim margin, and thus impose its will on forty-nine percent of the people. In a totalitarian society, change is brought about by revolution, and is generally abrupt. In a system where the majority rules, and the parties follow different policies, it is only a matter of time until one policy is replaced by the other, and the government see-saws along. Neither system is completely fair to the underdog. This realization is the main reason that Baha'is are forbidden to participate in any form of political activity.

The Baha'is avoid anything which could prevent their attainment of their goals of uniting mankind and establishing world peace. The Universal House of Justice states:

> If a Baha'i were to insist on his right to support a certain political party he could not deny the same degree of freedom to other believers. This would mean that within the ranks of the Faith, whose primary mission is to unite all men as one great family under God, there would be Baha'is who would be opposed to each other. Where, then, would be the example of unity and harmony which the world is seeking?[169]

It is true that the Baha'is' prohibition of participation in political activities will not save them from actual participation in politics in the future, should the administrative bodies of the Baha'i communities have reached such a level that such participation is inevitable and daily political matters must be resolved by the Baha'is.

In the Baha'i view, political administration cannot differ from the Baha'i administration, an institution which is already established and functioning smoothly. In the Baha'i administration there is no special person such as a president; the possibility of domination by any one person is eliminated. Instead there is an institution which is composed of nine duly elected members. They are elected without any form of campaigning, by a secret ballot of the members of the Baha'i community who are over twenty-one years of age. Even husbands and wives are not allowed to tell each other for whom they vote. Members are chosen for their ability to serve the community. The details of this unique administrative system are well-conceived, and extend from the local to national and international levels.[170]

Our concern now is whether a country dominated by Islam could transform into a nation of Baha'is, and how such a transformation could possibly take place.

To answer these questions, consider that it was only as long ago as 1971 when Ayatollah Khomeini first published his book *Hukumate Islami (Islamic Government)*. Who could have believed that only eight years later the mighty Shah would be deposed and that in his place would sit the unknown Khomeini? Who could have imagined that 2,500 years of Iranian monarchy would suddenly be tossed onto the ash-heap of history?

The main factors which could lead to a shift of power in favor of the Baha'is are the dismal circumstances suffered under the mullahs, an economy unable to provide the barest essentials operating in a political environment of terrorism, kidnaping and murder, along with general repression of the simplest human freedoms. In contrast, the Baha'i Faith offers a thoughtfully conceived plan for peace and freedom which would restore the economic health of Iran. The Baha'is are a great asset to Iran; all that remains is for the people to discover that fact.

Leo Tolstoy, referring to the Baha'i Faith, wrote:

> I believe that at this very hour the great revolution is beginning which has been preparing for two thousand years in the religious world–the revolution which will substitute for the corrupted religion, and the system of domination which proceeds therefrom, the true religion, the basis of which is equality among men, and the true liberty to which all beings endowed with reason aspire.

As Abdul-Baha defined "Baha'i": "To be a Baha'i is to become the embodiment of all human perfections." One can but hope that the Iranians will wake up and realize where their best interests lie.

12

IRAN AND ISRAEL:

Counterbalancing the Arabs

T he mutual hatred of the Israelis and the Arabs, and the constant threats to Israel of its Arab neighbors, are well-known. What is often forgotten is that the Iranians and the Arabs waste no love on each other, either.

Surrounded by the Arabs, in the unique geopolitical juxtaposition that fate has bestowed upon them, Iran and Israel's mutual interests include their very survival. Their common interests have led to a sort of marriage of convenience, and it is likely that this relationship will continue.

Certainly there is a lot of talk on the part of the mullahs about marching on Jerusalem. But this is just saber-rattling to enhance their political image and to keep the Israelis' respectful, lest they think that the threats from the Islamic radicals of Lebanon, Egypt, the occupied West Bank or Gaza Strip, inspired and financed by the Islamic Republic, are fading.[171]

The animosity between the Arabs and Iran dates back to the time of Mohammed, when, in 624 A.D., the Arabs invaded Iran and imposed Islam upon the Iranians. Ever since, the stage has been set for constant disputes and wars. Although the relationship has improved since the formation of OPEC, centuries of hostility have left indelible marks.

The disgust and contempt which the Iranians and the Arabs have for each other is evidenced in their languages. For example, in the Persian language, the word for Arab has a derogatory meaning; to call a person an Arab is a humiliating insult. The figure of speech "Are you an Arab?" degrades the person addressed to the level of a cow, and denotes low intelligence. For their part, the Arabs consider the Iranians to be *Ajams*, which means that they are non-Arabs, and barbarians. When applied to the Iranians it is an epithet of insult and ridicule.

The roots of the differences between Israel and the Arabs go even farther back in time, to the days of Moses and before. The Israelites and the Arabs descended from different sons of the patriarch Abraham, Isaac and Ishmael, who were born of different mothers.[172] From their sibling rivalry developed a tribal animosity which lasted through the time of Mohammed and up to the present.

The most recent source of hostility between the Arabs and the Israelis was the establishment in 1949 of the State of Israel in Palestine. With the occupation of Palestine and the displacement of the native Arabs from their homeland, Israel piled explosives on top of the tinderbox. Two wars, in June, 1967, and October, 1973, as well as constant skirmishes, terrorist bombings, hijackings and kidnapings, have overwhelmed the scenery of the Middle East and have practically become a way of life.

Until very recently, the common goal of all the Arab nations, and of the Palestinians in particular, was to retake Israel and drive the Jews into the sea. However, practical experience seems to have taught them that their goal is unattainable. Worse, with each attempt to destroy Israel, the Arabs lose more territory to the Jews. If it had not been for the political insight and diplomatic skills of the late Egyptian President Anwar Sadat, the Sinai would still be occupied by Israel, as the formerly Muslim part of Jerusalem and the Golan Heights are to this day.

As times have changed, so has the myth of invincibility that used to surround Israel. The bloody war between Iran and Iraq, and the introduction by Iraq of chemical weapons, have dispelled the myth somewhat. The vulnerability of Israel in the face of chemical warfare against its population is obvious, and the Israelis have reason to fear for their very existence. The Israelis have reason to be concerned about such weapons of mass destruction, in light of the small area of Israel, the capabilities of Arab nations, particularly Iraq and Syria, to produce chemical weapons, the Arabs' predisposition to destroy Israel, and their ability to deliver chemical weapons against Israel by missile.[173]

On the other hand, the usefulness of Israel's nuclear deterrent is practically obviated by the small size of Israel and the near-certainty of a backwash of nuclear fallout, Israel's limited ability to sustain casualties, the vast territory and resources of the Arabs, and the legitimate interests of the West in the Arabs' oil–and their marketplace.[174]

With the limited options available to the Israelis, it is not surprising that they decided to side with Iran and provide weapons to the Iranians in their war with Iraq. Although the economic motive of selling weapons to Iran cannot be overlooked, there is a more important strategic consideration which has tilted Israel toward Iran. This is the hostility of the Arab world, which dictates Israel's encouragement of regional factors which divert or absorb military and economic resources which could otherwise be used to exert pressure on Israel.[175]

While Israel faces the constant threat of invasion or outright destruction from all sides, it will have enough to do to simply maintain its existing territory; Israel's small population is declining. There has been a seventy-five percent reduction in immigration into Israel in recent years, during which time more people have left Israel than have entered. The exodus is blamed on disenchantment with Israel's political situation and high taxes, high inflation, high unemployment and compulsory military service.[176] With Israel unable to expand, the future spread of Zionism is at best a dream for its architects.

The present survival of Israel and Iran, surrounded by hostile Arabs, is dependent on their continued cooperation. By bringing together Israel's technology and Iran's population, these two opposing countries could reinforce their position as a counterbalance to the menace of the Arabs.

13

WINNING BACK IRAN

*I*f the goals of the West–particularly those of Great
Britain and the United States, in initiating and supporting Khomeini's revolution–were to destroy
Iran and its economy, to return Iran to the Middle Ages, to
prevent Iran from joining the developed world, then one
must certainly agree that these goals have been splendidly
achieved. But did their achievement bring the results desired
by the instigators?

To answer this question, one must consider the economic,
political and strategic reasons often presented as the motiva-
tions for bringing down the Shah and abetting Khomeini's
revolution. Consider also the gains and losses which have
accrued to both the Iranians and the West. There is no doubt
that the establishment of the Islamic Republic has brought
not the slightest benefit to the Iranians. Considering everything that has happened, it seems that the West didn't gain
much, either.

One motivation of the West in promoting the revolution in
Iran was Zbigniew Brzezinski's notion of blocking any Soviet
designs on the Middle East and Africa by interposing an
"Islamic Greenbelt".[177] The Muslims may not care much for
Americans, the reasoning went, but one thing they hate more
is communism.

The presence of the Shah, as both a staunch ally of the United States and a military power in his own right, probably would have done more to prevent the influence of the Soviets in the Middle East. Shortly after the fall of the Shah, the Soviets invaded Afghanistan, Israel's mythical invulnerability was debunked by the intifada movement, and the radical Arabs gained access to chemical weapons.

Most of the radical leaders in the Islamic Republic were trained in the Soviet Union or are sympathetic to the Soviets. Rafsanjani's first trip to any foreign country, shortly after Khomeini's death, was to the U.S.S.R., where he signed a series of long-term contracts with the Soviets, and reached an agreement of mutual non-interference in each others' domestic affairs. That is, the Islamic Republic would not stir up the Muslim population in the Soviet Union, and the Soviets would not oppose the mullahs' regime in Iran.

Considering only these facts, it is evident that the Islamic Greenbelt idea was a failure. Brzezinski, on one hand, was the only one in the Carter Administration who wanted to keep the Shah in power, while on the other he was the architect of the Islamic Greenbelt strategy; two completely contradictory notions. The results were that the dreaded communist menace was replaced by a far greater one: Islamic fundamentalism.

Inferences[178] that the Americans and the British wanted Iran's economy destroyed so that they could profit by reconstructing it lack credibility. If this is what the West wanted accomplished, it should have chosen someone other than the radical Muslims to bring it off. No one is more resistant to progress than the mullahs, and their adherence to the concepts of a "divine economy" precludes any meaningful expansion of the Iranian economy, least of all through involving foreigners or foreign capital. Others claimed that the world feared that Iran would become another Japan, and that the West promoted the revolution to prevent this. One must admit that the revolution certainly had a dampening effect on Iran's economic prospects.

So what benefit, if any, did the West derive from the establishment of a fanatical theocratic dictatorship in Iran? The only tangible result of the establishment of the Islamic Republic was a drastic drop in OPEC oil prices as the fundamentalists thumbed their noses at OPEC and pumped all the oil they could. Although this was a boon to most of the West, Great Britain and Norway, as oil-exporters, saw the prices of their oil drop, too.

The arms manufacturers and traders of the industrialized world enjoyed tremendous sales to both Iran and Iraq, and in reinforcing the other nations of the Middle East against their dangerous neighbors.

Weigh these benefits against the loss of one of America's strongest and most loyal allies.

Beyond the near-term negative effects of removing the Shah in favor of the fundamentalist Muslims, the West has taken a far greater risk in the Middle East. The consequences would be highly detrimental to the West, and the United States in particular. The most serious threat to the political and economic scene in the Middle East is the potential spread of the Islamic revolution beyond the borders of the Islamic Republic, especially to Egypt and Saudi Arabia. These two countries are the prime targets of the fundamentalists: Egypt for its large population, and Saudi Arabia for its oil wealth. Should these two countries fall to the Islamic revolution, the rest of the Middle East would inevitably follow. Egypt is vulnerable due to the extreme poverty of its people, and Saudi Arabia because of its small population, the high concentration of Shi'ites in the oil-rich areas of the country, and–in the eyes of the fundamentalists–the corrupt Saudi monarchy.

There is widespread belief that a Muslim Brotherhood, including all of the radical Islamic movements, is controlled by a British oligarchy whose intentions are to keep the nations of the Middle East in the Dark Ages, in order to continue their exploitation. It is said further that the plot is actually against Islam, but that the Muslim Brotherhood is an instrument of British interests and works to realize British policies. Because the fundamentalists get the opportunity to work for fundamentalist Islam, they are trapped as unwitting tools of their masters.[179]

The history of British colonial policy is full of examples which would support such a hypothesis. In retrospect it is hard to imagine Khomeini as a puppet of the West, unless all of his vituperation was for show. However, recent history is full of puppets who simply turned on their masters. The fact remains that one would rather be a hero to his own nation than remain a puppet of another. The following examples demonstrate the frailty of international puppet strings:

- Whatever the reasons the Germans had for sending Lenin to start a revolution in Czarist Russia, it is safe to assume that they did not intend for one half of Germany to end up dominated by Russia and the other half intimidated by her.

- Mahatma Gandhi led India, a colony of Great Britain, to independence.

- Zulfikar Ali Bhutto, the former president of Pakistan, failed to lead his country to independence, but not for want of trying.

- The Shah of Iran was returned to his throne by the United States after the overthrow of Dr. Mossadegh. The Shah tried to make Iran independent of economic and political influences from the West, without success. Interestingly the Shah, who became one of the most important leaders of the 1970s, got little attention from the superpowers in the early days of his rule. The Shah described the November, 1943, visit to Tehran of Franklin Roosevelt, Winston Churchill and Josef Stalin this way:

> Although I was technically the host of the conference, The Big Three paid me little notice. We were, after all, what the French called *quantite' negligible* in international affairs, and I was a king barely twenty-four years old. Neither Churchill nor Roosevelt bothered with international protocol that required they call on me, their host. Instead, I paid them courtesy visits in their embassy residences.[180]

Thirty years later the world came to recognize the Shah's significance, and to fear his increasing power.

- To the displeasure of the western world, Ayatollah Khomeini, placed in power by the British and the Americans, began to disobey the directives of his benefactors from the very beginning.

- Saddam Hussein is the newest of this breed of rulers.

It must be remembered that Japan and Germany rebuilt on the rubble of World War II and rose to become economic powers surpassed in the West only by the United States. Should the Islamic Republic manage to spread its revolution across the Middle East, the combination of the power of religion, the geostrategic position of the Middle East, the oil resources and production facilities–and the desperate dependence of the West on Middle East oil–and the potential to couple the flow of oil to monetary policy, could elevate Iran and the Middle East to the status of economic super-power.

If the Islamic fundamentalists were to gain control of the

Middle East, it is unlikely that even military action on the part of the United States could shake their hold. Consider the inability of the United States to effect the release of the hostages from its embassy in Tehran in 1980.[181] Remember the impotence of the Shah's army, one of the mightiest in the world, during the Iranian revolution.[182] This time, the foe is not a strong ruler or fanatical dictator; the foe is intangible: an ideology.

The competition for economic power among the industrialized nations of the West would probably work increasingly to the disadvantage of the Americans and the British. Since the Germans, French and Japanese, have no multinational oil companies nor any control over the flow of oil upon which their economies are strongly dependent, their economies have been largely dependent on the British and Americans, and their multinational oil companies. It is likely that Germany, France and Japan would begin to build a closer relationship with the oil producing nations of the Middle East in hope of gaining more control of the oil market–and their destinies.[183]

The following scenario illustrates how easily this could be accomplished. The French, Germans and Japanese could form an international oil consortium, which would function as their multinational oil company. This consortium would agree to make payment for oil half in the currency(ies) of the producers, and half in marks, francs and yen. They could further secure their oil supplies by providing technology to develop the economies of the oil-producing countries. Should the Japanese become a serious factor in the arms industry,[184] this would smooth the way even more.

It could turn out that the Germans, French and Japanese, with their multinational oil company, would become increasingly independent from the United States and Great Britain.

The oil-producing countries would become accordingly less dependent upon the British and Americans as well, and move toward consolidating their own financial power. The result would be a drastic drop in the value of the U.S. dollar and the British pound.

In addition to the possible threat from the Germans, French and Japanese, the interests of the West are also jeopardized by the efforts of the Soviets to gain control over the Middle East and its oil. Several agreements[185] between the Islamic Republic and the Soviets are evidence of these intentions.

While all these factors threaten to erode the position of the United States as the leading economic power in the world, American handling of foreign policy and its conduct of international diplomacy constantly undermine its own interests. The Americans are well known for their lack of political knowledge about their host countries and for their tendency to make shortsighted policies without regard for their long-term consequences. The Americans are famous for supporting dictators and other unpopular rulers. Americans do not seem to respect the people or customs of other nations, which generates a lot of hatred and resentment toward the United States.[186]

It is therefore imperative for U.S. interests that the United States establish long-term relationships with other countries based on mutual respect. This can be achieved only by supporting governments which have been freely and democratically chosen by the peoples of these countries, rather than regimes which have been imposed upon them. As much as the Islamic Republic would like the world to believe that it has the support of the majority of Iranians, there is no basis of truth to this. The very fact that the mullahs must use repressive and terroristic tactics to remain in power demonstrates that the majority of Iranians do not support them. In fact, the Islamic Republic is even less popular and more hated than was the regime of the Shah.

The very first prerequisite for the United States to establish a long-term peaceful relationship with the Iranians is to help them remove the brutal, repressive and unpopular theocracy of the Islamic Republic. Second, the United States must support a free election of a new government in Iran.

Is Iran worth the bother? The answer is yes, when the following factors are considered:

1)The removal of the Islamic theocracy in Iran would curb the spread of fundamentalist Islamic revolution in the Middle East. Fundamentalist Islam is inimical to the interests of the West, and of the United States in particular.

2)Iran has the potential to benefit the West and itself like no other country. Iran has tremendous oil, gas and mineral wealth. Its population of over fifty million deserve better than to live the way they do. They would gladly–and rightly–use their resources to provide a better life for themselves and future generations if they had creative leaders instead of repressive ones.

3)It is important to the whole world to prevent the creation of another heavily-populated and poverty-stricken country such as Egypt in the Middle East. According to the 1986 census,[187] the population of Iran was 49.8 million. This number increases by 3.5 percent annually, the highest rate of growth in the Third World. Every year 1,750,000 people are added to the existing population. The average age is seventeen years, which puts Iran among the countries with the youngest populations. Sixty percent of the population is fifteen or less; ten million are ten years old or less. By the year 2000, Iran's population will exceed 80 million. The age distribution in the population requires a drastic expansion of educational facilities and of the creation of jobs for the millions of new workers. The West can either ignore these developments and allow Iran to become another Pakistan, India or Egypt, or take the opportunity to convert Iran into a large and prosperous market.

4) Removal of the Islamic theocracy will remove the barriers to development caused by the Islamic economic concept.

5) A free and democratic Iran would dampen terrorist activities against Israel and the West.

So how could the removal of the mullahs be initiated, and by whom should they be replaced? To answer these questions, one must take a closer look at the personalities and political groups vying for power in Iran.

The first group is the monarchists, with 28-year-old Reza Pahlavi, the eldest son of the late Shah Mohammed Reza Pahlavi, as their potential king. Reza Pahlavi recalls to Iranians the era of peace and prosperity enjoyed under his father's rule; Reza Pahlavi would be seen to have inherited his father's throne, but not his regime. He is well-educated and sophisticated, and already has the support and sympathies of a broad base of political groups outside the Iranian government. Unlike his father, who was an absolute monarch, Reza Pahlavi claims that he would be a constitutional monarch, along the lines of King Juan Carlos I of Spain.[188] Reza Pahlavi's weaknesses are also his strengths: he is young and inexperienced, which also means that he has done nothing wrong and has no past to atone for.

Reza Pahlavi would be a viable candidate to lead a post-transitional government as king of Iran, should the Iranians decide to re-establish the monarchy. For the present, he lacks the means to return to Iran, unless by methods used by the CIA to return his father to the throne after the Mossadegh adventure. This would be the last thing that the United States would instigate now, and for that matter, the last way Reza Pahlavi would want to gain the throne.

The second group are the members of Dr. Mossadegh's National Front Party (Jabheh-e Melli). If Mossadegh was shortsighted in his abortive nationalization of Iran's oil industry, the present leaders of this party show no signs of greater competence to lead Iran into the future. The most prominent leaders of the National Front Party are Mehdi Bazargan, the first prime minister of the Islamic Republic, Dr. Sanjabi, the first foreign minister of the Islamic Republic, and Dr. Shahpur Bakhtiar, the last prime minister of the Shah's regime. Despite the total failure of these three leaders of the National Front Party during their time in power, the most disturbing facts are that Bazargan and Sanjabi ignored the Shah's pleas for help, openly took the side of Khomeini, contributed actively to Khomeini's rise to power and participated in the cabinet of his subsequent regime. Bakhtiar accepted the prime ministry at a critical time, but there is evidence to suggest that he did so in order to smooth the path for Khomeini's ascent to power.[189]

Clearly, it would be a mistake to imagine that these people could do any better this time. The National Front's "liberal" course is too irrational to make appropriate political and economic decisions. It began with Mossadegh's nationalization of the oil industry and ended up with Bakhtiar, Bazargan and Sanjabi handing the country over to Khomeini.

The third group vying for power in Iran is the Mujahidin-e-Khalq (Fighters for the People), whose background was presented earlier. Ervand Abrahamian, in his book *The Iranian Mujahidin*, best describes what would happen if the Iranian Mujahidins came to power:

Rajavi's personality cult had two far-reaching consequences. In the first place, it frightened off many former allies. If
the Mujahidin, these allies asked themselves, did not have a
semblance of democracy within their own organization, what
faith could be put in their promise to respect the political
rights of other organizations? If they were already, before the
revolution, worshipping their leader as a demi-god, what
type of personality cult would they create afterwards? If they
were using Shii imagery to legitimize their leaders' power,
what confidence could others have that their state would
separate religion from politics? If the Mujahidin in exile were
denouncing their critics, even sympathetic ones, as "traitors",
"parasites", "leeches", "garbage" and "gutter filth", how
would they deal with these adversaries once in power? In the
words of Hajj-Sayyed-Javadi: "With the triumph of the personality cult, the Mujahidin began to see the world in simple
black and white terms. Those who accepted the cult were
considered absolutely good. Those who refused were labeled
traitors, opportunists, and representatives of evil." Thus
many former supporters began to wonder in what way, if any,
the Mujahidin version of the Islamic Republic would differ
from that of Khomeini.[190]

The only difference would be that Khomeini's *Velayate
Faghih* would be replaced with Rajavi's Masuliyat,[191] exchanging one devil for another. This is not the sort of change
which Iran needs or wants. If the Mujahidin were to come to
power in Iran, Iran would be lost forever.

The Tudeh Party (Communist Party) is the fourth group.
The Tudeh Party has no political significance in Iran at the
present. In 1983, it fell into the bad graces of Khomeini's
regime, and its leaders were either killed, forced into hiding
or fled the country. A communist regime in Iran would serve
neither the interests of Iran nor those of the West.

Another individual who comes to mind is Abol-Hossein
Bani-Sadr, who was the first president of the Islamic
Republic. Since he lacks support of any kind among the

Iranians, his prospects in a   post-theocratic regime are slim to none. Perhaps his greatest usefulness would be as a media commentator on events in Iran.

The real hope for winning back Iran is the Iranian Army. The Iranian Army in the Shah's time was considered the sixth-ranked army in the world, with modern weapon systems, and well-trained officers and soldiers. When Khomeini came into power, the army suffered tremendously as over 5,000 very capable high-ranking officers and generals were executed, and thousands more were imprisoned or exiled. It seemed that the army was undergoing systematic extinction. Further weakening the army was the formation of the Pasdarans (Islamic Revolutionary Guard), which was established as a rival military organization subordinate only to its own commanders–and the mullahs–rather than to professional military personnel.

With the beginning of the Iran-Iraq war, the importance of the Shah's army was recognized, when it was desperately needed to counter the invasion by Iraq and to drive the Iraqis out of the occupied areas. Although much of the damage to the army had been irreversible, the remnants of the Shah's army possessed enough skills and nationalistic fervor to defeat Iraq, had the army been allowed to carry out its own strategies, rather than those of the irrational and un-knowledgeable mullahs.

Despite the massive executions of its officer corps, there are still many more brave, and now disillusioned, officers in the Iranian Army. The theocratic regime fears the army; there is no love lost on the part of the army, either. In 1980, shortly after the revolution, the Iranian army formed an organization named NEGAB (Mask), originally named the NVPA (Nezamiane Vatan Pareste Iran–the Patriotic Military Forces of Iran). This organization prepared a group of 900 well-trained and capable military officers for a special mission: to bomb Jambaran, the residence of Khomeini, take over all strategically important locations in Tehran, and end Khomeini's rule. Unfortunately, the plan was discovered, and before it could be executed, over 240 participants found themselves executed instead. It is believed that a foreign government revealed the existence of the plot to officials of the Khomeini regime.[192]

NEGAB, which also stands for Nejate Ghiyame Bozorge Iran (Rescue from the Big Insurrection in Iran) sought to install a constitutional monarchy in Tehran.[193]

In retrospect, it is truly tragic that NEGAB's efforts failed. Still, it demonstrated that there are enough patriotic officers in the army to remove the mullahs by means of a coup. Experienced generals are not required for this task; a handful of brave officers with the assurance that there would be no outside interference would be enough to topple the mullahs. When Reza Shah took over Iran and rescued it from foreign aggression, he was only a simple soldier, not even an officer. Iran's army has plenty of brave officers yet.

After a successful coup against the theocratic government, the army should form an interim government , clean up the country, establish law and order, and get the economy under some degree of control. The military should govern until the country is functioning normally again.

At that point, a referendum should take place in which the Iranians would decide whether they wished to establish a constitutional monarchy or a republic. Should they decide in favor of the constitutional monarchy, Reza Pahlavi would be the logical candidate to become the constitutional king of Iran. Should they decide in favor of a republic, a government should be formed by free democratic election.

In either case, the mullahs must be completely banned from any form of participation in the election process and from influencing its outcome in any way. The separation of the mullahs from any political function is so important that it should be the first point in the new constitution after the change of regime.

Two questions remain: which form of government would best serve the interests of Iran, and of the West; and whether the United States could help establish a freely elected government after the military had been in control.

There is no doubt that a strong democratic government in Iran would serve the interests of the Iranians and the West. However, Iran is a multiethnic nation, whose borders are shared with many different countries and cultures, and Iran has a great 2,500-year history of monarchy. Thus, a constitutional monarchy could be the better alternative for the Iranians, and equally beneficial to the interests of the West. A constitutional monarchy embodies the positive features of a republic. But a constitutional king, serving for life, would be a more integrating figure in the political scene than a president who serves for a few years. The new constitution could prevent the monarch from taking absolute power.

In the case of Iran, the United States would be in a strong position to influence the transition from a military to a civilian government. Iran's military strength is based on U.S. hardware, and the military would be predisposed to U.S. interests, in return for continued assistance, equipment and training.

Given the circumstances in Iran, a military coup seems the only way to initiate the changes to pull Iran back from the brink of destruction and to stabilize the turbulent, and critical, Middle East. It is time for the United States to support the establishment of a foreign government that has the support of the country's masses, rather than one that merely serves the short-term interests of the United States. Such a move in Iran would best serve the interests of Iran and the Middle East, as well as those of America.

14

STRATEGY TO BRING PEACE AND SANITY TO THE MIDDLE EAST

*T*he validity of U.S. Middle East policy to date is reflected in events in the region. That is, about what one could expect when regimes of thugs, terrorists and religious fanatics are put into power on the one hand, and strong support is given to kings and sheikhs, with little regard for the interests of their own people, on the other. This already explosive situation is exacerbated by America's devoted support of Israel, which contributes to make the United States the most hated country in the world in the eyes of the fanatic fundamentalist Muslims.

It is sad and embarrassing to observe the ignorance and shallowness of U.S. Middle East policy since the downfall of the Shah. Consider the events which have taken place in that time.

The American government didn't like the Shah's human rights record, so it was replaced by Khomeini. This did not turn out to be an improvement. While Khomeini's regime was intended to serve as an "Islamic Greenbelt" to block the spread of Soviet influence through the Middle East into

Africa, the Iranians ended up as kissing cousins with the Soviets. It could be said that the grand failure of Brzezinski's Greenbelt policy coincided with the grand failure of communism.

After recognizing its blunder with Khomeini, the United States, together with the other nations of the West, boosted Saddam Hussein as he waged war on Khomeini. Then he invaded Kuwait and became a threat to Western interests.

Although America has not yet resolved the Saddam situation, it has already started cozying up to Syria's President Hafez Assad. Not too long before the Iraqi invasion of Kuwait, America called Syria a "terrorist state." Despite Assad's sudden solidarity with America, and his pledge to place troops shoulder to shoulder with the American forces against Saddam Hussein, this again underlines the ineptitude and lack of principles in U.S. policy.

By asking for Iran's cooperation against Saddam, the United States tried to plant the seeds of another future debacle. Had Iran responded favorably, America would have overlooked a lot of Iranian mischief...until it was again too late.

Of course, politics does make strange bedfellows, and in a crisis situation one cannot be too picky about one's allies. The invasion of Kuwait certainly inspired a hasty alliance of convenience. However, this does not justify the shallowness of U.S. policy toward the Middle East. At the very least, it creates confusion and encourages negative developments, and helps to boost terrorism, fanaticism and radicalism in a region that already has more than its share of those commodities.

All of this would not be so crucial if it were not taking place in the Middle East, and if the stakes were not the vital oil supplies of the West, jeopardizing the economic well-being and political stability of the Western industrialized nations.

Now America faces the consequences of a decade of folly in its Middle Eastern policy and its virtual lack of an energy policy, which is why Middle East oil is so vital. American energy policy in the past decade could be called the "Stockman Doctrine," after President Reagan's first director of the Office of Management and Budget, David Stockman. Stockman branded the proposals for energy self-sufficiency of the Nixon, Carter and Ford administrations as "cramped, inward looking" strategies based on "Chicken Little logic", and advocated dependacy on the world market for energy. He dismissed concerns about potential OPEC extortion, and stated that only two policies were required: "strategic reserves and strategic forces."[194]

America's unwitting reliance on the Stockman Doctrine served it well during the 1980s, with abundant supplies of cheap oil glutting the market. Had the Reagan Administration acted to further drive down the price of oil, and then levied higher import taxes on it, the budget might have come closer to balancing. But, of course, this did not happen, and dependency on cheap Middle Eastern oil increased to record levels, right along with the budget deficit.

As long as supplies were plentiful, the Stockman Doctrine worked just fine. Then came Saddam Hussein's invasion of Kuwait, and the time came to resort to the "strategic forces", and the ineffectiveness and limitations of the Stockman Doctrine became all too clear.

As oil prices rose dramatically overnight, and the plodding deployment of military forces began, it became apparent that a critical share of the world's oil was under the control of a madman. And then came the realization that although the "strategic forces" might be capable of defeating Saddam Hussein's Third World army, that was no guarantee of the security of the oil supplies. The Iraqis might only be beaten after they destroyed Iraqi, Kuwaiti and Saudi oil fields.

It is clearly necessary that the supplies of Middle Eastern oil be secured and that threats to the interruption of that supply be eliminated. This can only be achieved by bringing peace and sanity to this troubled region. To accomplish this, nine objectives must be met:

1) Governments of military leaders and fanatic theocrats must be eliminated. Obviously, the most prominent of these are the governments of Iraq and Iran.

America and the West cannot afford to withdraw their forces from the Middle East until the fate of Saddam Hussein is resolved. This dictator and his army must be neutralized, together with its arsenals, its capabilities to produce and deliver chemical weapons, and any potential to develop nuclear devices. This is the first step toward establishing peace and sanity in the Middle East.

Regarding Iran, the most populous non-Arabic country in the Middle East must return to peace and prosperity. This would defuse many potential future threats to the region. With the removal of the theocratic regime in Iran, the wings of the radicals and terrorists would be clipped. A theocratic regime of the Islamic Republic of Iran cannot contribute to peace in the region as has been demonstrated in the last chapter.

2) International agreements must be made prohibiting the sale of sophisticated weapons, or providing weapons technology, to Middle East countries. Short-term profits from arms sales cannot justify the long-term insecurity of the world community. The region is too troubled with radicals and fanatics. The consequences of arming such people are, by now, well known.

3) A new, secular regime in Iran should establish a closer economic and political relationship with Israel. This would counterbalance the Arabs. America and the West should promote this goal.

4) The new regime in Iran must provide true freedom of religion. This would help defuse the radicalism and terrorism of Shi'ite Islam, and might ultimately lead to the de-Islamization of Iran. This would be in the interests of both the Iranians and the West alike.

5) Monarchies in Saudi Arabia, Kuwait and other Gulf countries should be protected, but the power of kings, amirs and sheikhs should be reduced from absolute to constitutional. More political and economic rights should be given to the people. This would help prevent future revolutions and would reduce radicalism.

6) Saudi Arabia and Israel should get special attention. Saudi Arabia's financial position should be strengthened by coupling that country's oil supplies to a monetary policy involving partial payment in Saudi currency, as explained earlier. This would reduce Saudi Arabia's vulnerability, as trouble in that country would reduce the value of its money. Those holding Saudi currency would have a vested interest in a stable Saudi Arabia. However, Saudi Arabia must not build up its military forces, and must remain dependent on Western military protection.

Israel, on the other hand, must remain militarily strong until the problems with the Palestinians are resolved. A weak Israel would be devastating to the interests of the West in the Middle East.

7) The best solution to the Palestinian situation would be to move the Palestinians to Jordan, and to prepare conditions to provide them with a better life, economically and politically. To provide the optimal political climate for this move,

Jordan's King Hussein should be reduced to a constitutional monarch. The Prime Minister of Jordan should be a Palestinian.

International cooperation should provide Western technology and financing from the oil-rich Arab nations to offer a better life to the Palestinian newcomers to Jordan. This would provide jobs, prosperity, political identity and a new homeland to the Palestinians. This would certainly be a far better situation than what they now have, or will ever have in current circumstances.

8) Syria must be democratized as well. This will be more easily accomplished now that the Soviets are leaning more toward the West, and their economy is in disarray. As a result, it would be much easier to finally establish peace in beleaguered Lebanon.

9) Simultaneously, vigorous efforts must be made to improve the election processes in all Middle Eastern countries. Political freedom must be expanded, including universal suffrage.

The implementation of these nine objectives would not be easy, but it would not be impossible. The problems in the Middle East are not isolated, but inseparably interrelated. By now it should be apparent that piecemeal solutions or simply reacting to events in this volatile region will just not do. Only a comprehensive solution, by addressing the needs of the people of the region, can finally bring peace and stability to the Middle East–and best serve the interests of the West, as well.

15

CONCLUSION

B oth times that Iran has tried to gain economic independence, it has gone to the brink of disaster. Each time crude oil played a dominant role, and each time Iran was foiled by the British. The first time was in the early 1950s, when Iran nationalized its oil industry, and the second time was in 1979, when the Shah was brought down by the revolution and Khomeini's Islamic Republic took power.

Given the shrewdness and finesse which brought down the Shah, the stupidity and shallowness of those who replaced him are doubly surprising. The British set a lot of machinery in motion to accomplish this: the mullahs, Iranian students, Amnesty International, the BBC, and others. But replacing the Shah with a gang of religious fanatics demonstrates that British cleverness was flawed by shortsightedness.

Whatever anyone imagined might be gained by returning Iran to the Middle Ages and threatening the entire region with revolution is hard to reconcile. The big losers, along with the Iranian people, were the industrialized nations of the West, whose interests were much better served by the Shah's regime. The only beneficiaries were Khomeini, the mullahs, and the ragtag Islamic Revolutionary Guard.

If President Jimmy Carter hoped that a religious regime would improve the human rights situation in Iran, he was sorely disappointed. If Zbigniew Brzezinski envisioned an Islamic greenbelt between the Soviets and the Middle East and Africa, he certainly did not expect that one day the new supreme religious leader *(Valiye Faghih)* of Iran would be a man educated in the Soviet Union. It is difficult to imagine that anyone could have hoped to benefit from the disintegration of the once-booming Iranian economy.

If the purpose of instigating the revolution in Iran was to bring down the price of oil, then this was accomplished–temporarily. If there were any hopes that this would buy time to develop alternate energy sources and free the West from its desperate dependence on imported oil, they were lost in the myopic euphoria that rose as the price of oil fell.

The momentum in the alternate energy industries which had been fueled by the sudden radical increase in the price of oil was lost. The once-burgeoning solar energy industry in the United States was virtually wiped out in early 1986, when the Reagan administration decided not to renew the expiring federal tax credits to purchasers of solar energy equipment. This decision to save a few short-term revenue dollars will prove to have been a very costly mistake for the United States if not the whole industrialized world.

It is ironic that, in an independent study, Amory Lovins' Rocky Mountain Institute concluded that the American government had spent no less than $46 billion to subsidize the nation's energy industry in fiscal 1984. Describing the expenditures in bang-for-the-buck terms to a House of Representatives subcommittee, research assistant Richard Heede said that nuclear power, which received $15.84 billion, supplied less than 100,000 BTU per dollar of subsidy. Comparably calculated, renewable energy sources provided several million BTU per subsidized dollar.[195] Killing the solar energy industry will doubtlessly be recorded in history as a colossal blunder.

Now, more than ever, the United States and the other industrialized nations must concentrate on the development of alternate, especially renewable, energy sources. Of all of the possible alternatives available, solar is the most promising. Although even with accelerated development solar energy would not provide a significant portion of the nation's energy requirements in the near future, it could do so within a decade.

Whatever the reasons for the Shah's downfall, it would be unfair to hold Great Britain and the United States solely responsible. The blame must be shared by a somewhat coordinated assortment of countries, institutions, organizations and individuals, who actively or passively contributed through lack of support, propaganda, demonstrations, ineptitude, animosity, treason, strikes, subversion and terrorist actions. Among these were France, West Germany, international human rights organizations, Khomeini, the religious fanatics, the clergy associated with the monarchy, the communists, the National Front, the Mujahidins, the Iranian human rights organizations, intellectuals, judges and lawyers, Prime Ministers Sharif-Emami, Azhari and Bakhtiar, Generals Fardust and Ghara-Baghi, Iranian women and students, the mass media, members of the Iranian parliament, the Palestinians, the Libyans, the workers and civil servants. The ignorance and illiteracy of the Iranian people were two critical facilitating factors. The Shah himself, in his blind complacency, contributed as well; perhaps he thought that the CIA would come to his rescue again.

The imposition of the regime of fundamentalist Shi'ite

Muslims began the reversal of decades of progress in Iran, and Iran's regression to an almost medieval state. The only purpose served by the mullahs was their rallying of the people to defend their country against the Iraqis, who would not have dared to attack the Shah. In light of the mullahs' blatantly inept conduct of the war, it might be supposed that the mullahs were in no hurry to end the war. As long as the Iranians were preoccupied with the Iraqis, they were less likely to concern themselves with the corruption and repression at home.

After ten years of war, poverty, executions, terrorism, destruction and brutal repression, what are the prospects for the post-Khomeini Islamic Republic of Iran? If the regime should last much longer and successfully spread it fanatical religious revolution throughout the Middle East, the Islamic Republic could well dominate the region, either through the direct seizure of territory or indirectly through its radical ideology. In such a position, the Islamic Republic could control a decisive portion of the world's oil supplies, thereby gaining enormous economic power to back its political clout. No one could expect the world to be any better off for that.

Whatever its merits as a religion, Islam's tenets and laws do not provide adequate guidance to solve the economic, political and social problems of the modern world, and thus can do nothing to help Iran to grow and prosper. Islam is more suitable as a pretext for conquest and destruction than as a foundation for peace. Lest there be any doubt about his priorities, Khomeini stated in 1979: "Some people have come to me and said that now the revolution is over we must preserve our infrastructure. But our people rose for Islam, not for economic infrastructure."

Had the Iranian people foreseen their fate, it is more likely that they would have arisen to defend the economic infrastructure from the mullahs.

For Iran to return to a path of peace and progress, and to rejoin the world economy, the rule of the mullahs must end. The stage must be set for close and friendly ties with Iran and a new regime which must enjoy the support of the majority of Iranians, not their fear. The sooner the United States takes steps in that direction, the sooner the threat of the spread of radical Islam throughout the Middle East can be eliminated, and peace can come to in that troubled part of the world.

Recommendations from various experts and scholars on Iran[196] that the United States seek normal relations with the "moderate" mullahs, such as those in Rafsanjani's government, are naive and ignore the realities of the situation. Any of the mullahs who might seek to normalize relations with the "Great Satan" would likely find themselves at the end of a hangman's rope as their more radical fellows cursed their treason. The durability of the limited supply of "moderates" in the government is doubtful.

Ridding Iran of the mullahs requires a bold move, most likely by the Iranian military forces. Once the country has been returned to a modicum of normalcy, it must be left to the Iranian people to decide on the type of government which will best serve their needs. A constitutional monarchy would offer the integrating figure of a king, while incorporating the democratic principles of a republic.

The flow of crude oil from OPEC should be coupled to a monetary policy. Implementation of such a policy by Saudi Arabia would re-establish the Saudis' position of leadership in OPEC, provide a stabilizing influence in the Middle East and secure an orderly flow of vital oil to the industrialized nations. This policy would increase the exchange rates of the Saudi dinar. The creation of a dinar bloc would, of course, devalue the U.S. dollar and the British pound, the present currencies of exchange in the oil market.

Such developments would dampen the spread of radical fundamentalist Islamic influence in the Middle East. The nations of the region, including Iran, would find the benefits of a dinar-based monetary system irresistible.

Despite rising oil costs and the devaluation of the dollar in the short term, the overall effect on the United States of such developments would be favorable. All the oil-importing countries of the West would face increasing expenditures for energy as the dollar and the pound devalued. However, West Germany and Japan would be much more adversely affected, due to their near-total dependency on imported oil. The United States would find itself more competitive in world markets, almost by default, and less vulnerable at home to Japanese and German imports. U.S. trade deficits would shrink, or even disappear.

As time passes, the Iranian people should take a dispassionate look at the progressive teachings and concepts of the Baha'i Faith, which could transform Iran into a progressive, free society worthy of a proud position among the family of nations.

Of course, in the short run, there is Saddam Hussein and the Iraqi army with which to deal. At least this is a tangible situation and, fortunately, Saddam Hussein does not represent a long-term threat.

Meanwhile, Iran is a ticking bomb. If nothing is done to defuse the fanatical theocracy of the mullahs, it is only a matter of time before the bomb detonates, destroying the Middle East and crippling the industrialized West.

Chapter Notes

I. THE TICKING BOMB

1. Alan Reynolds, "Unjustified Hysteria," Forbes, September 3, 1990; and Holman Jerkins, Jr., "Oil Marketplace Freedom Staving Off Big Sortages," *Insight on the News*, August 27, 1990.

II. SADDAM HUSSEIN: THE LESSER EVIL?

2. Casper W. Weinberger, "The Military Option", *Forbes*, September 3, 1990.

III. SHAH MOHAMMED REZA PAHLAVI: A LESSON FROM THE PAST

3. Mohammed Reza Pahlavi, the Shah of Iran: *Answer to History*. New York: Stein and Day, 1980. p. 70.

4. Christopher Tugendhat: *Gigant Erdoel, Wirtschaft, Politik, Strategie*. Vienna: Verlag Fritz Molden, 1968. p. 145.

5. Pahlavi, *op. cit.*

6. Ashraf Pahlavi: *Faces in a Mirror*. Englewood Cliffs, NJ: Prentice-Hall, 1980. pp. 134–144.

7. Mohammed Reza Pahlavi, *op. cit.*, p. 91; also, Amin Saikal: The Rise and Fall of the Shah. Princeton, NJ: Princeton University Press, 1980. p. 44 f.

8. Saikal, *op. cit.*, p. 67 f.

9. Mohammed Reza Pahlavi, *op. cit.*, p. 161 f.

10. *Ibid.*, p. 152.

11. Amir Taheri: *Chomeini und die Islamische Revolution*. Hamburg: Hoffman und Campe, 1985. pp. 147–148.

12. Mohammed Reza Pahlavi, *op. cit.*, p. 152 f.

13. *Ibid.*, p. 27.

14.Helmut Schmidt: *Menschen und Maechte*. Berlin: Siedler Verlag, 1987. p. 243.

15.Information about Freemasonry is from: William J. Whalen: *Handbook of Secret Organizations*; Milwaukee, 1966. pp. 46–66.; Stephen Knight: *The Brotherhood–The Secret World of the Freemasons*; Briarcliff Manor, NY, 1984; Esmail Raein: *Faramushkhaneh va Framasunari dar Iran (Masonic Lodges and Freemasonry in Iran)*; first edition, Italy, 1968, second edition, Iran, 1968, third edition, Great Britain, 1978); Amir Nejat: Secret Organizations and Freemasonry, Persian edition; San Mateo, CA: Eastern Publishing Society, undated.

16.Mohammed Reza Pahlavi, *op. cit.*, p. 160.

17.Nejat, *op. cit.*, p. 448.

18.R. Pashutan: *Harakat ba Jabr* (Move With Force); Bonn: Entesharate Zartosht, undated. p. 16 ff.

19.Mohammed Reza Pahlavi, *op. cit.*, p. 74.

20.*Ibid.*, p. 145.

21.Pashutan, *op. cit.*, p. 162 f.

22.Mohammed Reza Pahlavi, *op. cit.*, p. 108.

23.The author was among the first group to serve in the Iranian Education Corps in 1963, which was part of the first principle of the White Revolution. For fourteen months I taught reading, writing and arithmetic to the children of a northern Iranian village called Liavol-e-Olya in Rudbar-e-Rasht. Not only did my two colleagues and I have to pay part of our meager salaries of $30 per month to our supervisor, but we also had to give him the money which we received from the government for improvements to the newly-build schools. When I returned to this village in 1971, I was disgusted to find that the school I had built with the villagers was being used as a storage place for hay. The public bath that we had built was also destroyed. The only note of encouragement from my fourteen months' service there was that one of my pupils had finished high school and was planning further education. Two of my other sixty pupils had made it to the sixth grade.

24.Mohammed Reza Pahlavi, *op. cit.*, p. 22.

25.U.S. Congress, House Committee on International Relations Subcommittee on International Organizations, Hearings on Human Rights in Iran, August 3 and September 6, 1976, p. 7.

26.Harald Irnberger: *SAVAK oder der Folterfreund des Westens;* Hamburg: Rowohlt Verlag, 1977. p. 63 f.

27.*Ibid.*, p. 52 f.

28.*Ibid.*, p. 46.

29.According to Dreyfuss: It is fairly common knowledge that Amnesty International is a front for British Intelligence. At the top are those who know it for certain: RAmsey Clark, Sean McBride, and Conor Cruise O'Brien. An Amnesty advisor, Princeton's Richard Falk, wrote the section of the 1980s project devoted to human rights.

Amnesty's 1976 report alleged that the Shah's secret police had tortured and killed political dissidents; its purpose was to foster a climate across the globe in which the Iranian regime was viewed as barbaric and inhuman.

(Robert Dreyfuss: Hostage to Khomeini; New York: New Benjamin Franklin House Publishing Co., 1980. p. 22.

30.Mohammed Reza Pahlavi, *op. cit.*, p. 158.

31.*Kayhan*, London, No. 172, October 29, 1987. p. 7.

32.Pashutan, *op. cit.*, p. 215 ff. According to Pashutan, his information about Fardust is first-hand. Pashutan was employed by the Imperial Inspection Commission for over five years, and worked under Fardust.

33.Mohammed Reza Pahlavi, *op. cit.*, p. 57 ff.

34.Fereydoun Hoveyda: *The Fall of the Shah;* New York: Wyndham Books, 1980. p. 95 f.

35.Mohammed Reza Pahlavi, *op. cit.*, p. 127.

36.Ulrich Tilger: *Umbruch in Iran, Augenzeugenberichte, Analysen, Dokumente*; Hamburg, Rowohlt Verlag, 1979. p. 171 ff.

37.The palaces of the Shah were: The Marmar, the Golestan, The Niavaran, The Saad-Abad, the Malakabad in Mashad, the Aram in Shiraz, the Ramsar in Ramsar, and the Noushar Palace in

Noushar. It has been reported that the entrance gate of another, Farah-Abad, cost almost $600,000.

38. The universities from which the Shah received concessions were: the Technical Institute of Tehran, Technical University of Isfahan, Pahlavi University in Shiraz, the University of Farah, and the University of Reza Shah the Great.

39. Hoveyda, op. cit., p. 98 ff; and Saikal, *op. cit.*, p. 154 ff.

40. Henry Kissinger: *White House Years*; Boston: Little, Brown and Co., 1979. p. 1264.

41. Praynay Gupte: "Rhetoric and Reality in the Iranian Arms Trade," *Forbes*, October 19, 1987. p. 32 ff.

42. *Keyhan*, London, No. 177, December 3, 1987, p. 13.

43. Hoveyda, *op. cit.*, p. 96 f.

44. Mohammed Reza Pahlavi, *op. cit.*, p. 57 ff.

45. General Nassiri, the former head of SAVAK, was treated in the same manner by the Shah in the end, and suffered the same fate as Hoveyda.

46. Hovyeda, *op. cit.*, p. 208.

47. Mohammed Reza Pahlavi, *op. cit.*, p. 185.

48. Taheri, *op. cit.*, p. 194.

49. Mohammed Reza Pahlavi, *op. cit.*, p. 172.

50. It has been said that Ramsey Clark had developed an animosity toward the Shah due to the Shah's refusal to appoint Clark's law firm to represent the Pahlavi family in legal matters in the U.S. The Shah had hired the law firm of John McCloy in 1970 instead. McCloy's firm had a key role in sparking the hostage crisis: he successfully lobbied the Carter administration to admit the Shah to the United States for cancer treatment in 1979. After the Shah's death, Clark's firm represented the new Islamic Republic of Iran. Since Clark could not get what he wanted from the Shah, it is said that he turned completely against the Shah and began to guide and help the opposition. He is said to have met Khomeini long before the revolution, and to have arranged for the Al Fattah Palestinian group to cooperate with Knomeini. Clark was

often seen in Paris during Khomeini's sojourn there. (See Pashutan, p. 152 ff.) According to H. Ross Perot, who had gone to Iran shortly after the revolution to rescue two of his employees, he had seen Clark moving around freely in Tehran's prison as if he was orchestrating events. (See Ken Follett: On Wings of Eagles; New York: New Amrican Library, 1983; and Robert Dreyfuss: Hostage to Khomeini, op. cit.)

51.Schmidt, *op. cit.*, p. 232. Suprisingly, Schmidt does not discuss the progress of the Guadeloupe summit as regards the Iran affair, other than to mention it as one of the subjects to be discussed and decided upon at this meeting.

52.For details about Huyser's contacts, negotiations and actions during this exploit, see General Robert E. Huyser: Mission to Tehran; London: Andre Deutsch, 1986. For a thorough critique of the failure of Huyser's mission, see Assad Homayoun: "Iran: Into Harm's Way," *Global Affairs*, Vol II, No. 3, 1987, pp. 177–183.

53.General Ghara-Baghi: *Esrare Mamuriyate General Huyser dar bohrane Iran (Secrets of General Huyser's Mission in Iran's Crisis)*; Los Angeles: Maverick, 1989.

54.Huyser, *op. cit.*

55.Kissinger, *op. cit.*, p. 1258.

IV.KHOMEINI AND THE ISLAMIC REPUBLIC OF IRAN: THE RESULTS OF IGNORANCE

56.Ruhollah Mussawi Khomeini: *Tuzih-al-Massael (The Interpretation of Problems)*; Tehran: Ketabkhaneh Melli, 1979. p. 487.

57.*Ayatollah Khomeini Principles, Politiques, Philosophiques, Sociaux et Riligieux*, Jean-Marie Xaviers, translator; Paris, 1979, is one example.

58.*Playboy Report: Ayatollah Khomeini, Meine Worte, Weisheiten, Warnungen*, Weisungen; Munich: Moewig Verlag, 1979.

59.General Abbas Ghara-Baghi: Hagayeghi dar baraye Bohrane Iran (Facts About the Crisis in Iran); Paris: Sazamaane Chap va Entesharate Soheil, 1987. According to the response from a group of former officers of the Shah's armed forces, Ghara-Baghi's books, and especially his presentation of events during the crisis, are mostly wrong and misrepresent the facts. These officers allege that Ghara-Baghi's sole intention in writing his books was to exonerate his traitorous behavior during the crisis. See: Tahlili as Ketabe haghayeghi dar baraye Bohrane Iran, Peymanshekan, Beghalame Goruhi as Afsarane Arteshe Shahin-shahiye Iran (Analysis of the Facts About the Crisis in Iran–Guilty of Perjury, Written by a Group of Officers of the Shah's Armed Forces in Iran); undated, no location given.

The author has also noted inaccuracies and lack of credible research by Ghara-Baghi in writing his book. Some items presented as fact by him do not stand up under any scrutiny. As an example I cite this representative case. On pages 90 - 91 of his book, Ghara-Baghi writes: "...as the newspaper *Etelaat* reported when the officials of the army went to the house of General Ali Mohammed Khademi, the former managing Director of Iran Air, in order to arrest him, the general committed suicide." This statement with no supporting evidence does not do service to the accomplishments of General Khademi both in his managment of Iran Air as well as in his Presidency of International Air Transport Association (IATA). I reject Ghara-Baghi's statement based on the following facts as provided by various people with first-hand experience of the circumstances of his death as well as members of his family:

- Beginning with three days prior to his death there were published reports in at least two Tehran dailies, including Paygham-e Iran, that General Khademi had committed suicide! This makes the suicide story at least suspicious. (According to his son, Monib Khademi, who was in Tehran at the time they started getting phone calls from friends and associates for condolence after these published reports while General Khademi was alive!

- According to the coroners reports conducted at Tajrish Hospital and later the examination by his family physician, Dr. Nadji, the fatal bullet wound entered his brain from the back of the head in a way that would have been impossible for him to reach.

- According to eye witnesses the person who assassinated him entered his residence with a group of people, over-powered the Air Force privates who were present, before reaching him in a lower level room in his residence.

- The day after his assassination the judicial department of the Armed Forces (Dadrasani-e Artesh) began a formal investigation into his assasination. This case was later abandoned due to the disintegration of this department of the military because of the Iranian Revolution.

- Two days after this incident SAVAK arrested the Air Force privates who were present at his residence at the time because their version of the story did not match SAVAK's version. General Khademi's brother-in-law, Dr. Moayad, was arrested within a few days for communicating to the United Press International the family's statement of what had happened. This was later reported by the Newsweek magazine (November, 1978).

- Mrs. Khademi was arrested by SAVAK within two weeks for not agreeing to lie about what had happened. She was visited by the Red Cross representatives from Geneva while under arrest. She was later released.

- Finally the Revolutionary Committee of Shimiran did its own investigation after the revolution and confirmed the family's statement of what had happened.

While this is just one example that I have dwelled on I have used it to demonstrate lack of proper research and investigation by Ghara-Baghi.

60. *Saltanat va Masuliat (Monarchy and Responsibility)*; Toronto: Sarbazane, Iran, 1982.

61. The number of Baha'is executed by the Khomeini regime is over 300. They were given the choice of renouncing their faith or being executed. When they refused, they were accused of spying for the United States or Israel, and were summarily killed. The Baha'is were, of course, simply practicing their religion and refused to recant.

62. According to a publication of the Mujahidin, in 1986 over 12,000 Mujahids were executed, most of them less than 20 years old. By now the number must be much higher.

63.Abolhassen Bani-Sadr: *Khiyanat be Omid (Betrayal of Hope)*; no publisher cited, 1983.

64.Shahpour Bakhtiar: *Yekrangi (One and All of the Same Color)*, Persian edition; Mission Viejo, CA: Shadow Copy and Printing, 1982.

65.Pashutan, *op. cit.*, p. 163 ff.

66.Taheri, *op. cit.*, p. 189.

V. THE STRUGGLE TO CONSOLIDATE POWER

67.James A. Bill: "Resurgent Islam in the Persian Gulf," *Foreign Affairs*, Vol. 63, No. 1, Fall, 1984, p. 120.

68.Schmidt, *op. cit.*, p. 242.

69.Bill, *op. cit.*, p. 120.

VI. THE DESTRUCTIVE POWER OF THE ISLAMIC SHI'ITE

70.Although the books cited in this chapter represent only a small number of such authenticated publications, the interested reader can get these and many more dealing with similar topics at any Iranian book store in the Western countries, e.g. in Los Angeles, California.

71.To avoid appearances of prejudice, only sources from authors who are self-described devout or former Muslims have been relied upon as background for this chapter.

72.The biographical sketch of Mohammed was adapted from Dr. Roushangar: *Bazshenasi Qoran (Re-evaluation of the Koran)*; San Francisco: Entesharate Pars, 1985. pp. 18-38.

73.It has been said that Mohammed was killed by a Jewish woman who put poison in his food. See Ali Mir Fetrous: *Islam Shenasi (Recognizing Islam)*; 1983. p. 65.

74.Ali Dashti: *Twenty-Three Years* (no publisher or date cited). p. 191.

75.*Ibid.*, p. 275 ff.

76.H. M. Balyuzi: *Mohammed and the Course of Islam*; Oxford: George Ronald, 1976. p. 168.

77.Suroosh Irfani: *Revolutionary Islam in Iran, Popular Liberation or Religious Dictatorship*; London: Zed Books Ltd., 1983. p. 10.

78.*Ibid.*, p. 11.

79.For detailed information the reader may refer to the following compendium: *Shojah-ul-Dien Shafa, Touzih-ul-Masael, Pasokhaii be Porseshaye Hezar Saleh dar baraye Tasheyohe Din va Tasheyohe Dokondarane Din (Interpretation of Problems, Some Answered Questions, Since a Thousand Years, about the Shi'ites as a Religion and as a Source of Business)*, first edition, Paris. This 960-page work contains solutions to every problem even the most misguided person could imagine.

80.The formal clerical structure in the Shi'ite faith is as follows:

–Ayatollah-ol-Ozma (Grand Ayatollah), a Mujtahid as well as a Marja-e-Taqlid (Source of Imitation).

–Ayatollah, a Mujtahid but not a source of imitation.

–Hujat-ul-Islam (Manifestation of Islam), a Mujtahid who, unlike an Ayatollah, has not written a Resaleh (epistle) and has not specialized in any branch of Islam yet.

–Further down the line are tens of thousands of tallabe (religious students) and pishnamaz (prayer leaders).

81.Ruhullah Massawi Khomeini: *Hukumate Islami (Islamic Government)*; Tehran, 1971. p. 184.

82.Taheri, *op. cit.*, p. 149. However, according to others, Taheri's characterization of Ann Lambdon in the matter of Khomeini's book is inaccurate. See Michael Curtis, "Resurgent Islam," *Global Affairs*, Vol. II, No. 4, Fall, 1987. p. 185.

83.Irfani, *op. cit.*, p. 4.

84.Mohammed H. Habibi: *The Development of Religion and Ideology from Mazdak to Mojahed*; Washington: Iran Times newspaper, 1982. p. 39 ff.

85.Nooruddin Kianoori as related to Ayatollah Khomeini through his grandfather, Sheikh Fazurallah Noori, who was Khomeini's mother's uncle. See *Fedayeyane Islam*; Los Angeles: Left Forum, 1982. p. 50.

86.*Confessions of the Central Cadre of the Tudeh Party*; Tehran: Islamic Propagation Organization, 1983.

87.Habibi, *op. cit.*, p. 42.

88.Irfani, *op. cit.*, p. 12.

89.Mouvement Iran Libre: *Khiyanatha va Jenayate Mujahedin-e-Khalg (The Treachery and Crimes of the Mujahedin-e-Khalg)*; Paris: undated. p 10.

90.Irfani, *op. cit.*, pp. 116-148.

91.Ezatollah Homayunfar, "Doshman Shenacy, Dictatoriye Mujahedin-Mahjuni as Dictatoriye Akundi va Dictatoriye Russi Ast" ("Recognizing the Enemy–The Dictatorship of the Mujahedin is an Electuary of the Mullahs' Dictatorship and the Russian Dictatorship"), Keyhan, No. 190, March 10, 1988, London.

VII. IRAN AND OPEC

92.*Oeldorado* ii, published by ESSO, A.G., Hamburg, 1989.

93.*Ibid.*

94.*Ibid.*

95.*Ibid.*

96.*OPEC Annual Statistical Bulletin 1987*, published by the Secretariat of the Organization of Petroleum Exporting Countries, Vienna, 1988. pp. 78-87.

97.*Oeldorado 88* and *Opec Annual Statistical Bulletin.*

98.*Oeldorado 88.*

99.*Ibid.*, and the author's own calculations.

100.*Ibid.*, and the author's own calculations.

101.*Ibid.*, and the author's own calculations.

102."Strange Bedfellows in Vienna–OPEC Finds New Friends Who Also Want to Raise Prices," *Time*, May 9, 1988.

103.For a detailed analysis of this subject, see Farid Akhtarekhavari, "Prerequisites for an Efficient Substitution for Crude Oil," *Ekonomska Analiza*, Beograd, 1976, p. 276 ff; and *Die Oelpreispolitik der OPEC-Laender, Grenzen, Gruende und Hintergruende, Munich*: Weltforum Verlag, 1975. p. 78 ff.

104. *Oeldorado 88.*

105. In the early 1970s the author recognized the limited potential of nuclear energy and wrote a book on the subject (see note 103, above). At the present, the early euhoria surronding alternate energy sources is gone, nothing has happened, and the problem is at least as big now as it was then.

106. See note 105, above.

107. In 1973, U.S. President Nixon announced the commencement of Project Independence, designed to bring about America's independence from foreign oil by 1980. In Mr. Nixon's opinion, Project Independence would have historical significance no less than that of the Manhattan Project, which produced the atomic bomb, and the Apollo Project, which landed men on the moon. Project Independence never got off the ground, and was a total failure.

108. "Financing Project Independence: Tough Questions Demand Answers," *Commerce Today*, No. 25, Septermber 16, 1974.

109. A. Parker: "Living With Oil at $10 per Barrel," *Challenge*, February, 1975.

VIII. RTUNITY FOR SUPREMACY

110. Farid Akhtarekhavari: *Die OPEC im weltwirtschaftlichen Spannungsfeld, Ein Beitrag zur Discussion um die "neue Weltwirtschaftsordnung"*; Munich: Weltforum Verlag, 1976.

111. Apart from the grey market, there would be two ways for the U.S. to convert dollars to tomans: either exchanging them in Iran, or in the United States at Iran corresponding bank. Either way, the Federal Reserve would pay for the gains from exchange rate revaluations.

112. On March 17, 1975, the author warned the Shah of Iran about the limitations of the oil pricing policy and presented him with a concept similar to the one outlined here. The author received a thank-you letter from Jamshid Amouzegar, wishing him all the best. Had anyone in power had the foresight to realize the significance of this concept in those days, and attempted to utilize it in OPEC, the Middle East would be a very different place today.

IX. THE LEGACY OF KHOMEINI

113"Montazeri as zabanhaye ham-dourahayesh" ("Montazeri From the Viewpoint of His Classmates), Keyhan, London, April 6, 1989.

114.*Ibid.*

115.*Ibid.*

116.*Ibid.*

117.*Ibid.*

118."Ayatollah Montazeri as yek dide motafavet" ("Ayatollah Montazeri From Another Viewpoint"), *Keyhan*, London, April 6, 1989.

119.*Ibid.*

120.*Ibid.*

121.*Ibid.*

122."Montazeri goft: Nameii nevashtam ke khab ra as cheshme Imam bebarad" ("Montazeri Said: I Have Written a Letter Which Will Steal the Sleep From the Imam's Eye"), *Keyhan*, London, June 1, 1989.

123."Pesare Ayatollah Khomeini dar namehe sad-o-dah safheii khod be Ayatollah Montazeri nevesht: Agar daghigh amal nakonid, rafteid" ("Ayatollah Khomeini's Son Wrote in His 110-Page Letter to Ayatollah Montazeri: If You Don't Act Correctly, You Are Gone"), *Keyhan*, London, May 25, 1989.

124.The reprint of Ayatollah Montazeri's letter to Khomeini of March 22, 1989, appeared in Heyan, April 6, 1989.

125.Khomeini's reply to Montazeri's letter of resignation appeared in *Keyhan*, April 6, 1989.

126."Chetour cheraghe omre Khomeini khamush shod? Wa cheguneh Majlis Khobregan jaye khaliye ura por kard?" ("How Did the Light of Khomeini's Life Go Out? And How Did the Parliament Fill His Empty Place?") *Rouzegar-e-Now*, San Francisco, CA, and Vincennes, France, June/July, 1989, p. 7.

127.*Ibid.*, p. 8.

128."Ahmad Khomeini nemehe pedarash jaal kard" ("Ahmad Khomeini Forged His Father's Letter"), *Keyhan*, London, August 3, 1989.

129."Khameneh-i va maselehe valayete faghih" ("Khameneh-i and the Question of Valayete Faghih"), *Rouzegar-e-Now*, June/July, 1989, p. 31.

130.*Ibid.*, p. 30.

131."Iran's Factional Fights Make Business Outlook Poor," *Early Warning*, Washington, D.C., No. 3/89, March, 1989.

132."Rafsanjani Denies Power Struggle Under Way in Iran," Miami Herald, June 9, 1989; and "Iran Advances Vote on President and Constitution," *Christian Science Monitor*, June 12, 1989.

133.Amir Taheri: "Nur fuer den Uebergang?" *Die Welt*, Hamburg. No. 129, June 6, 1989.

134.Hilda Kindinger: "Nach zehn Jahren Brot fuer die Seele klagen die Perser mehr Wohlstand ein," Die Welt, Hamburg, No. 182, August 8, 1989; and Tony Walker: "A New Iran?", *World Press Review*, New York, August, 1989.

135."Rafsanjani Wins Iran Vote," Christian Science Monitor, July 31, 1989.

136."Didare Rafsanjani as Shouravi" (Rafsanjani's Visit to the Soviet Union), *Rouzegar-e-Now*, June/July, 1989, p. 39.

137."Taswirike Rafsanjani as siyasthaye dakheli wa khareji Jomhouriye Islami tarsim kard" ("The Picture Which Rafsanjani Drew of the Domestic and Foreign Policy of the Islamic Republic"), *Rouzegar-e-Now*, June/July, 1989, pp. 35-38.

138.Louise Lief: "Breaking the Ice on Assets, Iran's Need to Rebuild its Economy Gives Bush a Chance to Deal," *U.S. News and World Report*, August 28/September 4, 1989. Facts and figures were adapted from this article.

139."Jange ghodrat miyane Rafsanjani wa Mohtashemi" ("The Struggle for Power Between Rafsanjani and Mohtashemi"), *Keyhan*, August 17, 1989.

140."Ba kenar gosashtane Musavi, Mohtashemi va Ray-Shari, Rafsanjani ba mokhalefate tondruha ruberu khahad shod" ("By

Removing Musavi, Mohteshemi and Ray-Shari, Rafsanjani Will Face the Challenge of the Extremists"), *Keyhan*, August 24, 1989.

141. "Jange ghodrat miyane Rafsanjani..." *Keyhan*, August 17, 1989.

142. "Noure chesmi, Haj Ahmad Aghag" (Darling, Mr. Haj Ahmad), *Rouzegar-e-Now*, June/July, 1989, p. 32 ff.

143. "Vasiyatnamehe Khomeini" ("Khomeini's Testament"), Keyhan, June 15 and June 22, 1989. and "Vasiyanahmeye jabeja shodahe Khomeini" ("Khomeini's Replaced Testament"), *Rouzegar-e-Now*, June/July, 1989, p. 34.

144. Khomeini's poetry was in his own handwriting and was printed in *Rouzegar-e-Now*, June/July, 1989, p. 34.

X. THE ISLAMIC ECONOMIC CONCEPT: A SOLUTION?

145. Kindinger, *op. cit.*

146. "Dar faselehe salhaye 1358 ta payane 1367, jamiat Iran bish as 19 million tan afzayesh yafte ast" ("Between the Years 1358 [1979] and 1367 [1988] the Population of Iran Has Increased by Over 19 Million People"), *Keyhan*, August 24, 1989.

147. Kindinger, *op. cit.*

148. "Jange hasht saleh ba Iraq baraye mardome Iran 627 milliard dollar hazineh dashte ast" ("The Eight Years of War With Iraq Has Cost the Iranian People $627 Billion"), *Keyhan*, August 24, 1989.

149. Muhhamed Nejatullah Siddiqi: Muslim Economic Thinking, *A Survey of Contemporary Literature; Leicester, England: The Islamic Foundation*, p. 71

150. *Ibid.*

151. *Ibid.*

152. See, among others: al-Lababidi: *Islamic Economics, A Comparative Study*, Lahore, Pakistan: Islamic Publication Ltd., 1980; M.N. Siddiqui: *Some Aspects of the Islamic Economy*; Lahore, Islamic Publication Ltd., 1978; Monzer Kahf: *The Islamic Economy, Analytical Study of the Functioning of the Islamic Economic System*, Plainfield, IN: The Muslim Students' Association of the United States and Canada, 1978.

153."Shariat" means religious law.

154."Sunnah" means basic teachings from Islam coming from the *Koran*, the Prophet and Hedith; also included are the interpretations of the Koran and Hedith.

155.Siddiqui, op. cit., p. 124.

156.Abdul Quader Shaikh: "Zakat and Taxation, Problem of Equity and Justice," in: *Outlines of Islamic Economics, Proceedings of the first symposium on the economics of Islam in North America (Indianapolis, IN*, March, 1977).

157.See, among others: Siddiqi, M.N.: *Some Aspects of the Islamic Economy* (Lahore, Pakistan: Islamic Publication Ltd., 1978); Al-Lababidi; Monzer Kahf; and *Contemporary Aspects of Economic Thinking in Islam, Proceedings of the Third East Coast Regional Conference of the Muslim Students' Association of the United States and Canada*, Brentwood, MD: American Trust Publications, 1980.

158.Siddiqui, *Some Aspects of the Islamic Economy*, p. 124 f.

159.Siddiqui, *Muslim Economic Thinking...*, p. 63.

160.Ibid., p. 63 f.

XI.THE AVENUE TO A STABLE AND PEACEFUL IRAN

161.For detailed information concerning the systematic accusations and harassment of the Iranian Baha'is by the mullahs, see *The Baha'is and Iran;*; Luxembourg: Horizonte, 1986.

162.William Sears: *The Baha'is in Iran: A Cry From the Heart;* Oxford: George Ronald, 1982. p. 219.

163.*Yaddashtaye Kiniaz Dalgourki ya Asrare Peidayeshe mazhabe Bab wa Baha dar Iran (Diary of Kiniaz Dalgourki, or the Secret of the Emergence of the Religion of the Bab and Baha in Iran);* Tehran: Hafez Bookstore, undated.

164.Ahmad Kasrawi: *Bahaigari (Baha'ism);* Tehran: Chape Marde Amrouz, 1956. Kasrawi was terrorized by Navab Safavi, a member of the radical Fedayiyane Islam, after Kasrawi published his book on the Shi'ites.

165.Abdolhussein Navaii: *Fatneye Bab (The Sedition of the Bab);*

Tehran: Entesharate Bab, 1983; and Javad: *Baha'i che miguyad? (What Does the Baha'i Say?);* New York, 1967. Most surprising is Dreyfuss' presentation of the Baha'i Faith *(Hostage to Khomeini,* pp. 115-125), in which the information is mostly wrong or at least inaccurate and incomplete. It is beyond the scope of this book to correct each of these mistakes or allegations. It seems that Dreyfuss was obsessed with presenting every possible link with the British conspiracy for world domination without realizing that in so doing he contradicted himself and defeated the purpose of his book. Besides his inaccuracies regarding the Baha'is, Dreyfuss also misstates the names of Iranian politicians, locations, events, for example.

166. Total separation of religion and politics is nonexistent. "In God We Trust" emblazoned on U.S. currency is clearly mixing religion and politics, although the U.S. Constitution prohibits it.

167. This part of the history of the Baha'i Faith is adapted from Firuz Kazemzadeh: "The Baha'i Faith," a summary reprinted from the *Encyclopedia Britannica;* Wilmette, IL, 1974.

168. Interested readers can get more information by writing to The Baha'i National Center, 112 Linden Ave., Wilmette, IL 60091 U.S.A., or by contacting the Baha'i Local Spiritual Assembly listed in their local telephone directory.

169. *Political Non-Involvement and Obedience to Government, A Compilation of Some of the Messages of the Guardian and the Universal House of Justice;* compiled by Peter J. Khan; Australia, 1980. p. 14 f.

170. See Shoghi Effendi: *The Goal of a New World Order;* Wilmette, IL: Baha'i Publishing Trust; and *An Institution of the Baha'i Administrative Order,* compiled by the Universal House of Justice; Haifa, Israel, 1972; and *The Constitution of the Universal House of Justice,* Haifa, Israel, 1972. There are many other works available dealing with this subject in great detail.

XII. IRAN AND ISRAEL: COUNTERBALANCING THE ARABS

171. Fearing Wider Islamic Extremism, Israel Rethinks its Tilt Toward Iran," *International Herald Tribune,* Paris, No. 32555, October 26, 1987.

172. Abraham had a total of eight sons. Ishmael was born of his Egyptian wife, Hajar. Isaac's mother was Sarah. The other six were born of Katura, whom Abraham married after Sarah died.

173. Abigdor Haselkorn: "Arab-Israeli Conflict: Implications of Mass Destruction Weapons," Global Affairs, Fall, 1987. p. 135.

174. Yohanan Romati: "Israel and Nuclear Deterrence," Global Affairs, Spring, 1988. p. 175 f.

175. Yohanan Romati: "Israel and the Iraq-Iran Conflict: A Perspective," Global Affairs, Fall, 1987. p. 135.

176. World Press Review, May, 1988, p. 4.

XIII. WINNING BACK IRAN

177. See also Dreyfuss, op.cit., p. 5 ff.

178. Ibid., p. 9 ff.

179. Ibid., p. 99 ff.

180. Mohammed Reza Pahlavi, op. cit., p. 72.

181. Robert Dreyfus explains the U.S. raid's failure to release the hostages as follows:

> The administration's explanation for the raid's failure–that it had been caused by the simultaneous failure of three of the eight helicopters used–was simply not believed. According to many reports, the real reason for the debacle of the U.S. action, in which a helicopter and a huge C-130 air transport plane reportedly collided on the ground in Iran and burst into flames while trying to flee, was Soviet military intervention. One source said that the raid failed when an overflight of Soviet Mig-21s staged a show of force directly above the American landing party, and the commander of the raiding force then decided on a hasty retreat, leading to panic and the crash. Other sources with CIA connections reported that the U.S.S.R. had bombed the U.S. force almost as soon as it landed at the staging ground for Phase II of the raid, and that the administration's official version of the story was a coverup.

(Dreyfuss, op. cit. p. 68.)

182. One should not overlook the traitorour elements involved, which contributed to the total collapse of the Shah's army.

183. According to Dreyfuss:

> In 1978, the governments of France and West Germany led the European Community—with the single exception of Great Britain—in the formation of the European Monetary System, conceived, as one West German official put it, as a "seed crystal for the replacement of the International Monetary Fund." The EMS and its "Phase Two" European Monetary Fund embodied a program that challenged the "controlled disintegration" scenario of the Carter administration at every point, calling for the strengthening of the U.S. dollar, a return to the gold standard, expansion of nuclear energy production around the globe, and the revitalization of the industries of the advanced sector through an ambitious high-technology export program to industrialize the underdeveloped sector.

> The success of the new monetary system hinges on forging an alliance for developoment with the OPEC nations. As early as 1977, France and West Germany had begun exploring the possibility of concretizing a deal with the oil-producing countries in which Western Europe would supply high-technology exports to the OPEC countries in exchange for long-term oil supply contracts at a stable price. In turn, OPEC countries would deposit their enormous financial surpluses in Western European banks, and, eventually, into EMS's own institutions, which would then relend them to other countries in the Third World. With those credits, the underdeveloped countries could begin to gain access to European high-technology exports.

> When London discovered that it could not dissuade President Giscard-d'Estaing and West German Chancellor Helmut Schmidt from the EMS project in 1978—using ordinary deterrents—the green light was given to the Muslim Brotherhood to speed the destabilizatin of Iran.

> The chief countries of Western Europe, along with

Japan, are almost totally dependent upon their oil supply from the Persian Gulf, and in 1978 that supply came
from five states: Iran, Saudi Arabia, Iraq, Kuwait, and
the United Arab Emirates. By bringing down the Shah
and spreading chaos throughout the Middle East, the
Anglo-Americans calculated that they could knock out
Europe with the threat or actuality of an oil cut-off.

(Dreyfuss, op. cit., p. 13 f.)

184.Robert Neff and others: "Tokyo Wants its Arsenal Made in
Japan," Business Week, September 25, 1989.

185."Iran and the Soviet Union Have Fruitful Talks, " Insight on
the News, June 26, 1989. p. 36.

186.James Fallows: "South Korea Only Wants a Little Respect," U.S.
News and World Report, August 14, 1989.

187."Durnemaye Egtesadi va Maliye Iran" ("The Prospects of
Iran's Economics and Finance"), Rouzegar-e-Now,
January/February, 1989, No. 84, p. 87 f.

188.Charlotte Hays: "Renewed Hopes for Return to Iran," Insight,
July 3, 1989.

189."Anghelabe Iran dar pisgahe tarikh, Aya Bakhtiar mikhast
regime nejat bedahad?" ("Iran's Revolution in Front of History,
Did Bakhtiar Want to Rescue the Regime?"), Rouzegar-e-Now,
August/September, 1989, p. 47 ff.

190.Ervand Abrahamian: The Iranian Mojahedin; New Haven and
London: Yale University Press, 1989. p. 255.

191.Ibid., p. 253. Note: "Masuliyat" means "responsibility" in Farsi
(Persian).

192."Ghiyame 18 Tirma 1359 cheguneh boniyan geraft va
cheguneh be khun neshast?" ("How Did the Rise of 13 Tirma
1359 Take Shape and How Did it End Up in Blood?"), Keyhan,
June 13, 1989.

193.Ibid.

XIV. STRATEGY TO BRING PEACE AND SANITY TO THE MIDDLE EAST

194. Curtis Moore and S. David Freeman, "Energy Independence: We Have the Technology, Do We Have the Will?", *Miami Herald*, October 14, 1990.

XV. CONCLUSION

195. Bill D'Alessandro: "Dark Days For Solar–Will We Regret Shelving the Alternative Energy Source When Fuel Prices Rise Again?" *Eastern Review*, New York, January, 1989.

196. Assad Homayoun and Ralph Ostrich, "Post-Khomeini Iran," *Global Affairs*, Fall, 1989, No. 4, pp. 188-191.

Bibliography

BOOKS

Abrahamian, Ervand, *The Iranian Mojahedin* (New Haven and London: Yale University Press, 1989).

Akhtarekhavari, Farid, *Die Oelpreispolitik der OPEC-Laender, Grenzen, Gruende und Hintergruende,* (Munich: Weltforum Verlag, 1975).

--------------- *Die OPEC im weltwirtschaftlichen Spannungsfeld, Ein Beitrag zur Discussion um die "neue Weltwirtschaftsordnung"* (Munich: Weltforum Verlag, 1976).

Baha'i and Iran, The (Luxembourg: Horizonte, 1986).

Bakhtiar, Shapour, *Yekrangi [One and All of the Same Color,* Persian edition] (Mission Veijo, CA: Shadow Copy and Printing, 1982).

Balyuzi, H.M., *Mohammed and the Course of Islam* (Oxford: George Ronald, 1976).

Bani-Sadr, Abolhassen, *Khiyanat be Omid [Betrayal of Hope]* (no publisher cited, 1983).

Constitution of the Universal House of Justice, The (Hertfordshire, England: Broadwater Press, 1972).

Contempory Aspects of Economic Thinking in Islam, Proceedings of the Third East Coast Conference of the Muslim Students' Association of the United States and Canada (Brentwood, MD:

American Trust Publications, 1980).

Dashti, Ali, *Twenty-Three Years* (no publisher or date cited).

Dreyfuss, Robert, *Hostage to Khomeini* (New York: New Franklin House Publishing Co., 1980).

Efendi, Shoghi, *The Goal of a New World Order* (Wilmette, IL: Baha'i Publishing Trust, 1938).

Follett, Ken, *On Wings of Eagles* (New York: New American Library, 1983).

Ghara-Baghi, Abbas, *Hagayeghi dar baraye bohrane Iran [Facts About the Crisis in Iran]* (Paris: Sazamaane Chap va Entesharate Soheil, 1987).

Ghara-Baghi, Abbas: *Esrare mamuriyate General Huyser dar bohrane Iran [Secrets of General Huyser's Mission in Iran's Crisis]* (Los Angeles: Maverick, 1989).

Habibi,   Mohammed H., *The Development of Religion and Ideology from Mazdak to Mojahed,*   (Washington: Iran Times newspaper, 1982).

Hoveyda, Fereydoun, *The Fall of the Shah* (New York: Wyndham Books, 1980).

Huyser, Robert E., *Mission to Tehran* (London: Andre Deutsch Ltd., 1986).

Irfani, Suroosh: *Revolutionary Islam in Iran, Popular Liberation or Religious Dictatorship* (London: Zed Books Ltd., 1983).

Irnberger, Herald, *Savak oder Folterfreund des Westens* (Hamburg: Rowohlt Verlag, 1977).

Islamic Propagation Organization, The, *Confessions of the Central Cadre of the Tudeh Party* (Tehran, 1983).

Javad: *Baha'i che miguyad? [What Does the Baha'i Say?]* first and second editions (New York, 1967).

Kahf, Monzer, *The Islamic Economy, Analytical Study of the Functioning of the Islamic Economic System* (Plainfield, IN: The

Muslim Students; Association of the United States and Canada, 1978).

Kasrawi, Ahmad, *Bahaigari [Baha'ism]* (Tehran: Chape Marde Amrouz, 1956).

Kazemzadeh, Firuz, *The Baha'i Faith* [summary reprinted from Encyclopedia Britannica] (Wilmette, IL, 1974)

Khomeini, Ruhollah Mussawi, *Tuzih-al-Massael* [The Interpretation of Problems] (Tehran: Ketabkhaneh Melli, 1979).

----------*Velayate Faghih [The Government of the Theologian]* (Tehran: Entesharate Amir Kabir, 1981).

----------*Hukumate Islami [The Islamic Government]* (Tehran, 1971).

----------*Kashfe Asrar [The Unveiling of Secrets]* (Qom, Iran: Entesharate Azadi, no date).

Kissinger, Henry, *White House Years* (Boston: Little, Brown and Co., 1979).

Knight, Stephen, *The Brotherhood, The Secret World of the Freemasons* (Briarcliff Manor, NY, 1984).

Lababibi, Al, *Islamic Economics, A Comparative Study* (Lahore, Pakistan: Islamic Publications Ltd., 1980).

Mir Fitrous, Ali, *Islam Shenasi [Recognizing Islam]* (no publisher or date cited).

Khiyanatha va Jenayate Mujahedin-e-Khalg {The Treachery and Crimes of the Mujahedin-e-Khalg] (Paris: Mouvement Iran Libre, undated).

Navaii, Abdolhussein, *Fatneye Bab [The Sedition of the Bab]* (Tehran: Entesharate Babak, 1983).

Nejat, Amir, *Secret Organizations and Freemasonry*, Persian edition (San Mateo, CA: Eastern Publishing Society, undated).

Pahlavi, Ashraf, *Faces in a Mirror* (Englewood Cliffs, NJ: Prentice-Hall, 1980).

Pahlavi, Mohammed Reza, the Shah of Iran, *Answer to HIstory*

(New York: Stein and Day, 1980).

Pashutan, R, *Harakat ba Jabr [Move With Force]* (Bonn: En-tesharate Zartosht, undated).

Playboy Report: Ayatollah Khomeini, Meine Worte, Weisheiten, War-nungen, Weisungen (Munich: Moewig Verlag, 1979).

Raein, Esmail, *Faramushkhaneh va Framasunari dar Iran [Masonic Lodges and Freemasonry in Iran]* (first edition, Italy, 1968; second edition, Iran, 1968; third edition, Great Britain, 1978).

Roushangar, *Bazshenasi Qoran [Re-evaluation of the Koran]* (San Francisco: Entesharate Pars, 1985).

Saikal, Amin, *The Rise and Fall of the Shah* (Princeton, NJ: Prin-ceton University Press, 1980).

Saltanat wa Masuliat [Monarchy and Responsibility] (Totonto: Sar-bazane Iran, 1982).

Schmidt, Helmut, *Menschen und Maechte* (Berlin: Siedler Verlag, 1987).

Sears, William, *The Baha'is in Iran, A Cry From the Heart* (Oxford, Ronald George, 1982).

Shaikh, Abdul Quader, *Zakat and Taxation, Problem of Equity and Justice,* in: *Outlines of Islamic Economics,* Proceedings of the first symposium on the economics of Islam in North America (In-dianapolis, IN, March, 1977).

Siddiqi, Muhhamed Nejatullah, *Muslim Economic Thinking, A Survey of Contemporary Literature* (Leicester, England: The Is-lamic Foundation, 1981).

----------*Some Aspects of the Islamic Economy* (Lahore, Pakistan: Islamic Publication Ltd., 1978).

Taheri, Amir, *Chomeini und die Islamische Revolution* (Hamburg: Hoffmand und Campe, 1985).

Tilger, Ulrich, *Umbruch in Iran, Augenzeugenberichte, Analysen,* Strategien (Vienna: Verlag Fritz Molden, 1968).

Tugendhat, Christopher; Gigant Erdoel,Wirtschaft, Politik, Strategie. (Vienna: Verlag Fritz Molden, 1968).

Universal House of Justice, comp., *An Institution of the Baha'i Administration order* (Haifa, Israel, 1972).

U.S. Congress, House Committee on International Relations, Sub-committee on International Organizations Hearings on Human Rights in Iran, August 3 and September 6, 1976.

Wahlen, William J., *Handbook of Secret Organizations* (Milwaukee, 1966).

Yaddashthaye Kiniaz Dalgourki ya Asrare Peidayeshe Mazhabe Bab wa Baha dar Iran [Diary of Kianiaz Dalgourki or the Secret of the Emergence of the Religion of the Bab and Baha in Iran] (Tehran, no publisher or date cited).

ARTICLES

"Ahmad Khomeini namehe pedarash jaal kard" (Ahmad Khomeini Forged His Father's Letter), *Keyhan*, London, August 3, 1989.

Akhtarekhavari, Farid, "Prerequisites for an Efficient Substitution for Crude Oil," *Ekonomska Analiza*, Beograd, 1976.

"Anghelabe Iran dar pishgahe tarikh, Aya Bakhtiar mikhast regime nejat bedahad?" (Iran's Revolution in Front of History, Did Bakhtiar Want to Rescue the Regime?), *Rouzegar-e-Now*, Vincennes, France, August/September, 1989.

"Ayatollah Montazeri as yek dide motafavet" (Ayatollah Montazeri From Another Viewpoint), *Keyhan*, April 6, 1989.

"Ba kenar gosashtane Musavi, Mohtashemi va Ray-Shari, Rafsanjani ba mokhalefate tondruha ruberu khahad shod" (By Removing Musavi, Mohteshemi and Ray-Shari, Rafsanjani Will

Face the Challenge of the Extremists), *Keyhan*, August 24, 1989.

Bill, James A., "Resurgent Islam in the Persian Gulf," *Foreign Affairs*, Vol. 63, No. 1, Fall, 1984.

"Chetour cheraghe omre Khomeini khamush shod? Wa cheguneh Majlis Khobregan jaye khaliye ura por kard?" (How Did the Light of Khomeini's Life Go Out? And How Did the Parliament Fill His Empty Place?) *Rouzegar-e-Now*, June/July, 1989.

Curtis, Michael, "Resurgent Islam," *Global Affairs*, Vol. II, No. 4, Fall, 1987.

D'Alessandro, Bill, "Dark Days For Solar - Will We Regret Shelving the Alternative Energy Source When Fuel Prices Rise Again?" *Eastern Review*, January, 1988.

"Dar faselehe salhaye 1358 ta payane 1367, jamiat Iran bish as 19 million tan afzayesh yafte ast" (Between the Years 1358 [1979] and 1367 [1988] the Population of Iran Has Increased by Over 19 Million People), *Keyhan*, August 24, 1989.

"Didare Rafsanjani as Shouravi" (Rafsanjani's Visit to the Soviet Union), *Rouzegar-e-Now*, June/July, 1989.

"Durnemaye Egtesadi va Maliye Iran" (The Prospects of Iran's Economics and Finance) *Rouzegar-e-Now*, January/February, 1989.

Fallows, James, "South Korea Ony Wants a Little Respect, *U.S. News and World Report*, August 14, 1989.

"Fearing Wider Islamic Extremism, Israel Rethinks its Tilt Toward Iran," *International Herald Tribune*, Paris, No. 32555. October 26, 1987.

"Financing Project Independence: Tough Questions Demand Answers," *Commerce Today*, Washington, D.C., No. 25, Septermber 16, 1974.

"Ghiyame 18 Tirma 1359 cheguneh boniyan geraft va cheguneh be khun neshast?" (How Did the Rise of 13 Tirma 1359 Take Shape and How Didi it End Up in Blood?), *Keyhan*, June 13, 1989.

Golan, Galia, Gorbachev's Middle East Strategy, *Foreign Affairs*, Vol. 66, No. 1, Fall, 1987.

Gupta, Frenay, "Rhetoric and Reality in the Iranian Arms Trade," *Forbes*, October 19, 1987.

Haselkorn, Avigdor, "Arab-Israeli Conflict: Implications of Mass Destruction Weapons," *Global Affairs*, Winter, 1988.

Hays, Charlotte, "Renewed Hopes for Return to Iran," *Insight, on the News*, July 3, 1989.

Homayoun, Assad, "Iran Into Harm's Way," *Global Affairs*, Vol. II, No. 3, 1987.

Homayoun, Assad, and Ralph Ostrich, "Post-Khomeini Iran," *Global Affairs*, No. 4, Fall, 1989.

Homayunfar, Ezatollah, "Doshman Shenacy, Ditatoriye Mujahedin-Mahjuni as Dictatoriye Akundi va Dictatoriye Russi Ast" (Recognizing the Enemy - The Dictatorship of the Mujahedin is an Electuary of the Mullahs' Dictatorship and the Russian Dictatorship), *Keyhan*, March, 1988.

"Iran After Khomeini," *Global Affairs*, No. 4, Fall, 1989.

"Iran and the Soviet Union Have Fruitful Talks," *Insight, on the News*, June 26,1989.

"Iran's Factional Fights Make Business Outlook Poor," *Early Warning*, No. 3, March, 1989.

"Jange ghodrat miyane Rafsanjani wa Mohtashemi" (The Struggle for Power Between Rafsanjani and Mohtashemi), *Keyhan*, August 17, 1989.

"Jange hasht saleh ba Iraq baraye mardome Iran 627 milliard dollar hazineh dashte ast" (The Eight Years of War With Iraq Has Cost the Iranian People $627 Billion), *Keyhan*, August 24, 1989.

Jenkins, Holman, Jr., "Oil Marketplace Freedom Staving Off Big Shortages," *Insight on the News*, August 27, 1990.

"Khameneh-i va maselehe valayete faghih" (Khameneh-i and the Question of Valayete Faghih), *Rouzegar-e-Now*, June/July, 1989.

Kindinger, Hilda, "Nach zehn Jahren Brot fuer die Seele klagen die Perser mehr Wohlstand ein," *Die Welt*, Hamburg, No. 182, August 8, 1989.

Lief, Louise, "Breaking the Ice on Assets, Iran's Need to Rebuild its Economy Gives Bush a Chance to Deal," *U.S. News and World Report*, August 28/September 4, 1989.

"Montazeri as zabanhaye ham-dourahayesh" (Montazeri From the Viewpoint of His Classmates), *Keyhan*, April 6, 1989.

"Montazeri goft: Nameii nevashtam ke khab ra as cheshme Imam bebarad" (Montazeri Said: I Have Written a Letter Which Will Steal the Sleep From the Imam's Eye), *Keyhan*, June 1, 1989.

Moore, Curtis, and S. David Freeman, "Energy Independence: We Have the Technology, Do We Have the Will?", *Miami Herald*, October 14, 1990.

Neff, Robert, et al., "Tokyo Wants it Arsenal Made in Japan," *Business Week*, New York, September 25, 1989.

"Noure chesmi, Haj Ahmad Aghah" (Darling, Mr. Haj Ahmad), *Rouzegar-e-Now*, June/July, 1989.

Parker, A., "Living With Oil at $10 per Barrel," *Challenge*, January/February, 1975.

"Pesare Ayatollah Khomeini dar namehe sad-o-dah safheii khod be Ayatollah Montazeri nevesht: Agar daghigh amal nakonid, rafteid" (Ayatollah Khomeini's Son Wrote in His 110-Page Letter to Ayatollah Montazeri: If You Don't Act Correctly, You Are Gone), *Keyhan*, May 25, 1989.

"Rafsanjani Denies Power Struggle Underway in Iran, *Miami Herald*, June 9, 1989.

"Rafsanjani Wins Iran Vote," *Christian Science Monitor*, July 31, 1989.

Ramati, Yohanan, "Israel and Nuclear Deterrence," *Global Affairs*, Spring, 1988.

Ramati, Yohanan, "Israel and the Iraq-Iran Conflict: A Perspec-